José Luis Guevara Valdez
Daniel Díaz Plascencia
Diana González López

Apple yeast additive in growing tilapia diets

José Luis Guevara Valdez
Daniel Díaz Plascencia
Diana González López

Apple yeast additive in growing tilapia diets

Beneficial yeasts in the feeding of growing Tilapia (Oreochromis niloticus)

ScienciaScripts

Imprint

Any brand names and product names mentioned in this book are subject to trademark, brand or patent protection and are trademarks or registered trademarks of their respective holders. The use of brand names, product names, common names, trade names, product descriptions etc. even without a particular marking in this work is in no way to be construed to mean that such names may be regarded as unrestricted in respect of trademark and brand protection legislation and could thus be used by anyone.

Cover image: www.ingimage.com

This book is a translation from the original published under ISBN 978-3-659-08706-6.

Publisher:
Sciencia Scripts
is a trademark of
Dodo Books Indian Ocean Ltd. and OmniScriptum S.R.L publishing group

120 High Road, East Finchley, London, N2 9ED, United Kingdom
Str. Armeneasca 28/1, office 1, Chisinau MD-2012, Republic of Moldova, Europe
Printed at: see last page
ISBN: 978-620-7-73005-6

Contents

GENERAL SUMMARY

In the present study, several experiments were carried out to evaluate the effect of fermented apple bagasse and the use of a yeast-based additive in the diet of tilapia fish (*Oreochromis niloticus*) during the growth stage. Both the fermented bagasse and the yeast additive were prepared by aerobic fermentation processes. During the preparation of the yeast inoculum, the effect of two natural additives, Zeolite and Oregano Essential Oil (OEO), on the microbiological safety parameters was evaluated, where the treatment with OEO showed a considerable decrease of aerobic mesophilic bacteria. It did not differ significantly against the control in the yeast count at 48 h ($P > 0.05$). Another study evaluated the antimicrobial effect of AEO and zeolite on solid-state fermentation (SSF) of apple bagasse, where maximum yeast growth was obtained at 48 h with the addition of both treatments (462×10^6 cell/g), while the control achieved it at 96 h (470.5×10^6 cell/g), reducing the traditional fermentation time and optimising the process. The *in vivo* trial was carried out in a closed fish tank system, with controlled conditions of oxygenation, pH and water quality, for 8 weeks. Morphometric dimensions of the fish were taken weekly, and once the productive parameters of growing juvenile fish were determined, they were sacrificed to determine the fillet yield. It was found that fish fed the BMF had a higher weight of 37.93 g, while the control treatment had a weight of 34.92 g ($P > 0.05$). The yeast inoculum treatments were 32.42 g, being lower ($P>0.05$). Length gain was also a parameter that varied by treatment, with the BMF treatment being higher (92.19 mm) than the control (88.38 mm), and as with the other variables, the length gain of fish fed with the yeast inoculum was considerably lower (83.84 mm; $P>0.05$).

GENERAL INTRODUCTION

Aquaculture is the development of aquatic organisms under controlled or semi-controlled conditions, thus being one of the best economic options for the production of animal protein. The last decade has seen a development in aquaculture worldwide, although focusing on developing countries, as it has the potential to produce more fish at lower cost, covering the demand that will arise due to the demographic explosion without affecting the ecology and producing safe, high quality food with an excellent nutritional profile (FAO, 2020).

In aquaculture fish production, 50 % or more of the expenditure is directed to feed, so it is necessary to develop a complete, balanced, highly digestible feed with added nutritional benefits and high protein quality, as well as additives that preserve fish health such as nutraceutical and probiotic supplements in order to optimise fish growth and achieve a higher productive benefit (Thiessen *et al.*, 2003; Idenyi *et al.*, 2022).

When formulating a feed for aquacultured fish, more than half of the cost is directed towards adding protein of acceptable quality, with fishmeal usually being the main input used for this purpose. Fishmeal has an adequate balance of amino acids, fatty acids, digestible energy, vitamins and minerals, making it a very good choice for inclusion in fish diets (Shaeffer *et al.*, 2009; Hoyos *et al.*, 2017).

However, there are alternative protein sources available on the market that are derived from vegetable sources, such as soybean meal, which, having an excellent amino acid profile, represents an excellent opportunity to supplement fishmeal without having to direct so much economic resource to the protein source (Dong *et al.*, 2013).

Bagasse ferment represents a very interesting option when formulating a diet for fish produced by aquaculture, as it is the product of microbial fermentation of the by-products of the apple industry, it represents an economical alternative to add protein of microbial origin, with high digestibility and extremely accessible price (Balcazar *et al.*, 2006), as well as offering added value such as the presence of antioxidant compounds from the apple and yeasts from fermentation, which add their probiotic capacity to the diet (Diaz-Plascencia, 2011; D^az-Plascencia 2017a).

Therefore, the general objective of the present work was to evaluate the productive and meat quality parameters of tilapia (*Oreochromis niloticus*) fed a diet supplemented with solid fermented apple bagasse and a yeast inoculum produced during fermentation.

The particular objectives were to quantify the yeasts comprising the inoculum produced in this fermentation and to describe the relationships between the functional compounds and the health of fish fed with these compounds.

Aquaculture

The term aquaculture refers to the cultivation of aquatic organisms whose main destination is human consumption, under controlled conditions, which translates into a better productive yield and a lower economic investment. For more than two thousand years, the Chinese civilisation has been producing fish by raising them in aquaculture systems, both in ponds specifically made for this purpose and in flooded rice fields during the rainy season. Therefore, the history of aquaculture goes back several millennia in the history of mankind, who over the years has tried to obtain food from fish, but reared in limited bodies of water, in order to control the parameters necessary to obtain the maximum benefit (Wang *et al.*, 2008).

The rapid growth of the human population has led to the need to consider new food sources that contain high quality protein, a good lipid profile and, above all, are affordable for the general population. Aquaculture meets all these requirements, since, in a relatively small system, with minimal environmental pollution and with little effort on the part of the fish farmers, a food of high nutritional value for the consumer can be obtained at a very low cost. In order to sustain the high demand for aquaculture products for human consumption, the aquaculture industry has had to make adjustments to its methods and facilities, the most important of which is the intensification of cultivation (Wang *et al.*, 2007).

As demand for aquaculture products increases, the farmer is forced to diversify his crops, increase stocking density and become part of the commercial chain, bringing into play factors outside the normal water controls, such as parasites, viruses or bacteria, which enter the system through the introduction of new organisms or hatchlings from infected farms. All these infectious problems cause technical and economic problems, involving pond disinfection, drug costs and, above all, loss of sperm (Bondad-Reantaso *et al.,* 2005).

As Mexico is a country with extensive coastlines and diverse bodies of water, it is to be expected that fishing activity has a high importance with respect to other productive sectors, so that in 2011, 1,660,475 tonnes of live weight were obtained in fisheries, of which 262,855 tonnes were obtained from aquaculture systems, which had a share of 15.83 % of the national total (SAGARPA and CONAPESCA, 2012).

In 2011 Chihuahua ranked 27th among the states of the country in fish production with 758 tons of production, which represents 0.05 % of the national production. Chihuahua has aquaculture fish production centres, which contribute significantly to this production, with a main production of carp (308 tons), followed by trout (230 tons), mojarra (173 tons), catfish (113 tons) and lobster (14 tons), which makes it the tenth landlocked state with fish production (SAGARPA and CONAPESCA, 2012).

Aquaculture is therefore an activity of high socio-cultural importance, especially in rural environments, where small-scale producers can obtain good quality feed at very low prices, generate employment and diversify the productive practices of livestock producers, It also provides a feed with a better nutritional profile than any other terrestrial animal as it is a source of high quality animal protein, provides highly digestible energy and is an important source of omega-3 fatty acids, polyunsaturated fatty acids, fat-soluble vitamins and minerals (Bondad-Reantaso *et al.*, 2005).

Feeding in Aquaculture

Aquatic organisms, like all terrestrial animals, need nutrients such as proteins, carbohydrates and carbohydrates, as well as vitamins and minerals for their development, from which they obtain the energy and components necessary to carry out their metabolic functions. These nutrients come from the food they can find in their natural environment, such as plant detritus, other fish or organic matter that falls from terrestrial ecological systems, while in aquaculture systems they are limited to the diets provided by the aquaculturist, with the exception of semi-controlled systems, where they can also obtain food from natural sources to a lesser extent (Lovell, 1991).

Diets formulated for fish have higher protein requirements than diets for terrestrial animals, mainly because fish have higher protein requirements, but lower energy requirements. This leads to the higher cost of formulating a diet to include protein and should be formulated on the premise that the protein included is intended to generate protein for deposition in the fish, rather than for use as an energy source (Liu *et al.*, 2011).

Fish have the advantage of needing much less metabolisable energy than terrestrial animals, mainly because they consume less energy when moving in their environment, maintaining their posture or temperature, and because they excrete their nitrogenous waste in the form of ammonium, they consume less energy in the catabolism and excretion of protein waste (Gaylord *et al.*, 2009).

Starches are easily digested by fish, especially by warm-water fish, but are less readily utilised by freshwater fish, which tend to use primarily the catabolism of fatty acids and triglycerides for energy. This is why some freshwater fish, such as salmon, have an obligatory requirement for omega-3 and 6 fatty acids in their diet, while temperate water fish need triglycerides to carry out their lipid metabolism with a lower proportion of omega-3 and 6 fatty acids (Lovell, 1991). The same author reports that formulating a diet for aquaculturally reared fish, in which all the nutrients necessary for optimal development of the fish are provided, will have a high economic cost which leads to the search for new methodologies to facilitate the digestion or absorption of the compounds provided in the diet. This field of research in formulation leads to functional compounds in animal nutrition, which provide an additional benefit to their nutritional value.

Tilapia (*Oreochromis niloticus*) and its Cultivation

Tilapia is the ninth most farmed species worldwide, with Nile tilapia (*Oreochromis niloticus*), Java tilapia (*O. mossambicus*) and blue tilapia (*O. aureus*) being the most sought-after species. They are native to Africa and the Middle East, but have now been introduced to almost all tropical areas of the world, mainly as a species for farming in small ponds in rural areas and mainly for self-consumption by people in precarious situations. Tilapia is now a highly sought-after species, with important markets in Japan, the United States and Europe, as well as in some developing countries (Shaeffer *et al.*, 2009).

Tilapia are sensitive to temperature and have an optimum temperature of 28 to 32 °C. As the temperature decreases, the growth of the fish is reduced, until it stops eating when the temperature reaches 16 or 17 °C, and dies at less than 10 °C. They are freshwater species, but can grow at salinities up to 25 ppt (Shaeffer *et al.*, 2009).

The meat obtained from tilapia is a soft, light-flavoured, light grey or white flesh with some red colouration near the lateral line. In species that are fortified, the flesh is lighter in colour than with the pure species, and in adults over 600 grams the flavour is intensified, especially in the lateral line. This species has a fillet yield of approximately 33-35% of live weight, which is relatively low (Cnaani and Hulata, 2008). The same author mentions that the tolerance of tilapia to variations in environmental conditions, its high fecundity and ease of competition with other species have made it a highly sought-after species by aquaculturists, easy to grow and tasty.

Apple Bagasse

The state of Chihuahua is the first national apple producer, with 43.1 % of the national planted area and 73.7 % of the total production, producing 370,040.2 tons per year (SAGARPA, 2009), with which it can be estimated that a quarter of this is waste apple because it does not meet the quality standards. This waste, together with the by-products generated by the apple processing industry, generates waste, mainly apple bagasse, being 25 to 30 % of the weight of the fresh fruit, which represents a serious pollution problem, as it is highly biodegradable and significantly alters the environment in which it is discarded thanks to its high content in carbohydrates, acids, fibres, vitamin C and minerals (Joshi and Sandhu, 1996).

Thanks to the low nitrogen content and high moisture content of apple bagasse, its inclusion in animal diets has been difficult, a situation that makes it imperative to develop techniques that allow its consumption and reduce the environmental impact it generates. Therefore, one of the most successful alternatives is the fermentation in solid state (FES) of bagasse, added with urea nitrogen and mixtures of vitamins and minerals that allow the development of yeasts, which in addition to degrade the fibre, can provide protein of microbial origin, i.e., supplement at low cost high quality protein (Diaz *et al.*, 2010).

Functional Compounds in Animal Nutrition

Functional foods are those that contain a compound or nutrient with some beneficial activity, i.e. that can provide a physiological outcome in addition to their own nutritional value. Like most animals, fish are also susceptible to oxidative stress caused by the presence of free oxygen radicals, and although there is a whole enzymatic mechanism to capture the latter, they are not always sufficient to eliminate the radicals generated in the fish's tissues. The antioxidant enzyme system is mainly composed of the enzymes glutathione peroxidase, glutathione reductase, superoxide dismutase and catalase, as well as some reduced substances, such as glutathione. Oxidative stress occurs when this antioxidant enzyme system decreases, or the presence of free radicals is increased, which directly affects the health of the fish as cell membranes are attacked by free radicals and tissue damage occurs that can affect the development of the fish (Dong *et al.*, 2013).

To mitigate the generation of free radicals it is necessary to include in the diet antioxidant compounds that can help the enzyme system of the fish in the capture of free radicals, especially those coming from the oxidation of oils, since the toxicity of oxidised oils in the diet of fish is well documented (Yamashita *et al.*, 2009).

Among the compounds that can be added to a diet as antioxidants are polyphenols, which are natural constituents of plants, which use them as antioxidants against free radicals produced in photosynthesis (Biedrzycka and Amarowicz, 2008).

Although hundreds of polyphenols have been identified, the two most important types of polyphenols are flavonoids and phenolic acids, which in turn are divided into several classes. Flavonoids include flavones, flavonols, flavanols, flavanones, flavanones, isoflavones, anthocyanins, among others, and are commonly found in onions, tea, apples, citrus fruits, soybeans, berries and cocoa. And the most important phenolic acids are caffeic acid and ferulic acid, found in coffee and some vegetables (Scalbert and Manach, 2005).

In the case of apples, there are differences between varieties with regard to the quantity of polyphenols present, the most outstanding being Braeburn, Red Delicious, Cripps Pink and Granny Smith, Idared and Rome Beauty, with between 66.2 and 211.9 mg/100g fresh weight (depending on the variety), with the flavanols catechin and proanthocyanidins being the most present, followed by hydroxycinnamates. These flavonoids are synthesised by the plant in response to bacterial infection, as they exhibit antimicrobial activity by binding to bacterial membrane proteins and destabilising them (Biedrzycka and Amarowicz, 2008).

From the by-products of the apple, a food product called "*manzarine*" was produced, which was obtained by fermentation in a solid state, where the apple's own flora was developed using the carbohydrates of the apple itself and urea as a source of nitrogen. With this fermentation process, a protein product is obtained which, in addition to increasing the natural protein of the apple thanks to the proliferation of microorganisms, presents live yeasts, including *Saccharomyces cerevisiae*, *Kluyveromyces lactis* and *Issatchenkia orientalis* (Diaz-Plascencia, 2011; Balcazar *et al.*, 2013).

These yeasts have been studied in other species as probiotics, where improvements have been achieved (Diaz-Plascencia, 2011; D^az-Plascencia 2017a). It is therefore a point of interest to know their behaviour as Probiotics in acucola species, in order to evaluate their performance and their benefits to the host.

Probiotics in Animal Nutrition

With the intensification of aquaculture worldwide and the globalisation of trade in aquaculture species, a number of pathogenic bacteria have spread among ponds and aquaculture centres worldwide. This type of behaviour has led fish farmers to use antibiotics to control infections and improve the health status of the fish. However, the use of antibiotics leads to problems such as the emergence of antibiotic-resistant bacterial strains or the misuse of antibiotics, leading to the slaughter of fish that have not yet completely eliminated the antibiotic, or to overdosing. To avoid the problems associated with antibiotic use, alternative methods have been developed that are able to keep the levels of pathogenic bacteria low through the inoculation of beneficial bacteria that compete with them for substrate, enzyme attack or attachment site in the gut of the acucolous organisms. Having a well-controlled population of these beneficial bacteria significantly reduces pathogenic bacterial populations and stimulates the fish's immunity, making antibiotic use unnecessary (Bondad-Reantaso *et al.*, 2005).

Probiotics were originally described as living organisms that are added to a food and have a beneficial effect on the host through the intestinal balance they create. In addition to the original description, non-living microbiological components have been added and their area of action has been extended outside the host gut (Kesarcodiwatson and Kaspar, 2008).

In aquaculture, probiotics such as lactic acid bacteria, found naturally in the intestine of some fish, have been used successfully. These bacteria convert lactose into lactic acid, acidifying the intestinal environment and preventing colonisation by other pathogenic bacteria. Other bacteria studied are *Bacillus* spp, producers of bacteriocins, which attack pathogenic bacteria. Among the yeasts, the most studied is *Saccharomyces cerevisiae,* which has an immunostimulatory effect on the digestive tract of fish (Kesarcodiwatson and Kaspar, 2008).

In order to determine whether a yeast or bacterium is capable of being a probiotic, it must meet the following restrictions: 1) It must not be pathogenic to the host; 2) It must be accepted by the host and be able to replicate in the host; 3) It must be resistant enough to reach the site where it is intended to multiply; 4) It must be able to function *in vivo*, as well as *in vitro*; 5) It must not contain virulence or antibiotic resistance genes (Kesarcodiwatson and Kaspar, 2008).

Another benefit of using probiotics in aquaculture diets is that these microorganisms produce digestive enzymes, especially targeting compounds that fish have no way of digesting, such as complex carbohydrates, thereby aiding digestion by releasing smaller molecules from the enzymatic degradation of complex compounds, thereby increasing the digestibility of the feed and its utilisation by the fish (Wang, 2007).

Solid State Fermentation

Solid-state fermentation (SSF) is defined as the growth of microorganisms on solid material in the absence or near absence of free water, which has been used in recent years as an economical solution for the large-scale production of various organic metabolites (Bellon-Maurel *et al.*, 2003; D^az-Plascencia 2017b).

FES is currently used to produce compounds of industrial value, but in the agricultural area it has gained relevance thanks to the ease of fermenting agricultural waste at low cost and giving it added value for animal feed, which results in cleaner production processes, less aggressive to the environment and with a higher economic yield than traditional methods that generate tons of waste (Pandey, 2003).

Several factors affect the efficiency of FES, one of the most important being the nature of the substrate used, which is selected on the basis of its abundance and ease of procurement. The importance of the substrate is that in addition to providing the energetic substrates, it will provide sustenance for the microbial cells, so the shape and size of the particle is of vital importance for the proper development of the system. Small particle sizes provide more surface area for enzyme attack, while too small a particle size risks agglomeration and thus reduced microbial growth, while very large particles, although allowing adequate gas flow and little agglomeration, prevent enzymes from attacking the interior of the particle, thus wasting substrate by hindering mass transfer (Couto and Sanroman, 2006).

Another of the most important factors when carrying out an FES is aeration, which provides for the exchange of gases, mainly oxygen and carbon dioxide, adding the former and removing the latter, so that the process is carried out in an oxygen-rich environment. Aeration also serves to dissipate heat, distribute moisture throughout the system and remove volatile metabolites. By generating adequate aeration, the system will remain stable, as the pO2 and pCO2 values will remain homogeneous in the substrate and there will be no areas where changes in the aerobic conditions will modify the flora that develops there, thus avoiding contamination by organisms foreign to the FES. The moisture distribution will also allow for a homogeneous aw, so there will be no areas where conditions force microorganisms to modify their growth conditions (Graminha *et al.*, 2008; D^az-Plascencia 2017b).

Oregano Essential Oil

Oregano (*Lippia berlandieri*) is a wild plant from arid zones, belonging to the *Verbenaceae* family, which is used in traditional Mexican cuisine, in traditional medicine and, in recent years, in industry for the extraction of its essential oil (Navarrete *et al.*, 2005).

Extraction of Oregano Essential Oil (OEO) occurs by various methods, and is obtained because of its high content of antioxidant compounds, mainly monoterpenes such as thymol, carvacrol and terpinene and p-cymene, phenolic acids and flavonoids (Figure 1). Of these compounds, thymol and carvacrol are the main compounds in quantity (Table 1), as well as in microbial activity, which is well documented against bacteria such as *Escherichia coli*, *Staphylococcus aureus*, *Bacillus cereus*, *Vibrio* sp., *Clostridium* sp., and some fungi such as *Aspergillus*, *Penicillum* and *Geotrichum* (Renteria *et al.*, 2014).

The mode of action of essential oils has been extensively studied, as they are generally recognised as safe (GRAS) products, and their use in food and pharmaceutical products has become widespread, which requires a broad knowledge of their mode of action. In the case of AEO, this can be summarised in the ability of thymol and carvacrol to penetrate the bacterial cell membrane and generate a destabilisation, producing a defect in the proton motive force, mainly through the pH gradient (ApH) and the electric potential gradient (D y), This impedes the respiratory activity of the microorganisms and leads to leakage of ions towards the outside of the cell, thus generating their inactivation or death, depending on the type of membrane they have and the severity of the damage they receive due to their characteristics (Lambert *et al.*, 2001).

AEO has been used as a natural alternative to artificial preservatives, given its antioxidant activity, consumer safety, antioxidant capacity and, above all, its antimicrobial capacity (Teixeira *et al.*, 2013).

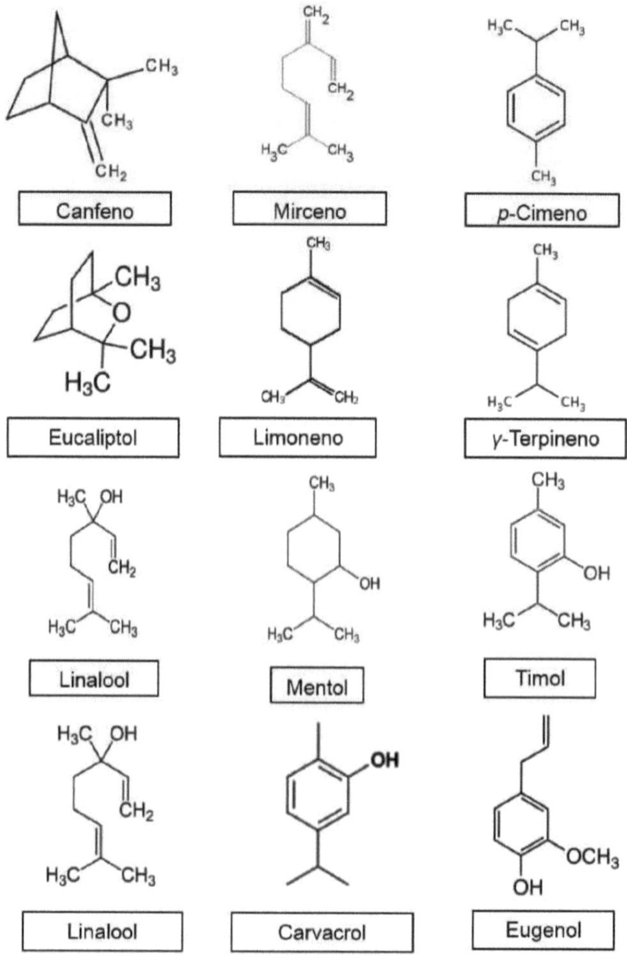

Molecular structure of the various compounds present in the essential oil of oregano (*Lippia berlandieri)* ecotype Jimenez, Chihuahua, Mexico (Silva, 2015).

Table 1. Compounds identified in the gas chromatography of oregano essential oil obtained from oregano plants from the region of Jimenez, Chihuahua (Silva, 2015).

Retention time (minutes)	Compound	% of Abundance
5.737	Camphene	0.0523
7.373	Myrcene	0.1256
7.925	p-Cimeno	16.1256
8.164	Eucalyptol	0.3185
8.226	Limonene	0.2568
9.284	y-Terpinene	5.3541
10.720	Linalool	0.2521
13.785	Menthol	0.7996
17.535	Thymol	3.3796
17.741	Carvacrol	60.0488
18.665	Eugenol	0.1357
19.476	Trans-Caryophyllene	3.6905
	26 other minority compounds	9.4608

Due to these properties, it is considered that the addition of AEO to apple bagasse FES will provide a defence against contamination by bacteria that will consume the substrates and decrease the fermentative efficiency of the yeasts.

Zeolite

Zeolites are clayey materials obtained from sedimentary deposits. Of the more than 40 types of naturally occurring zeolites, clinoptilolite is particularly functional in adsorbing nitrogen compounds, especially NH_4^+. Clinoptilolite is a type of zeolite that has an open molecular structure, with a pore volume of about 35 % and is easily found in mineral deposits worldwide. This characteristic is obtained thanks to the three-dimensional structure of aluminosilicates with four- and five-membered rings, thus forming micropores capable of harbouring exchangeable cations such as NH_4^+, Na^+, K^+ and Ca^{2+} (Leung et al., 2007).

Zeolites have been used for their ability to perform cation exchange between their structure and the medium in which they are dissolved. Their chemical structure $(Na_6[(Al_2O_3)(SiO_2)_{30}p24H_2O])$ allows them to adsorb cations, such as ammonium, and slowly release them depending on the concentration of the medium in which they are found, by concentration difference, thus modifying the pH of the medium and preventing the loss of nitrogen to the environment, which favours its use (Li, 2003).

LITERATURE CITED

Balzazar, J. L., I. de Blas, I. Ruiz-Zarzuela, D. Cunningham, D. Vendrell and J. L. Muzquiz. 2006. The role of probiotics in aquaculture. Vet. Microbiol. 114:173-186.

Bellon-Maurel, V., O. Orliac and P. Christen. 2003. Sensors and measurements in solid state fermentation: A review. Process Biochem. 38:881-896.

Biedrzycka, E. and R. Amarowicz. 2008. Diet and Health: Apple polyphenols as antioxidants. Food Rev. Int. 24:235-251.

Bondad-Reantaso, M. G., R. P. Subasinghe, J. R. Arthur, K. Ogawa, S. Chinabut, R. Adlard, Z. Tan and M. Shariff. 2005. Disease and health management in asian aquaculture. Vet. Parasitol. 132:249-72.

Cnaani, A. and G. Hulata. 2008. GenomeMapping and genomics in fishes and aquatic animals. GenomeMapping and Genomics in Animals. Vol. 2. p. 101 - 116.

Couto, S. R. and M. A. Sanroman. 2006. Application of solid-state fermentation to food industry-A review. J. Food Eng. 76:291-302.

D^az, D., F. Salvador, O. Ruiz, C. Arzola, A. Flores, O. La O and A. EKas. 2010. Effect of urea and soybean paste level on protein concentration during solid state fermentation of waste apple (*Malus domestica*). Revista Cubana de Ciecia Agricola 44 23-26.

D^az-Plascencia, D. 2011. Development of a yeast-based inoculum and its effect on in vitro fermentation kinetics in rations for high-producing Holstein cows. PhD thesis. Faculty of Zootechnics and Ecology. Autonomous University of Chihuahua. Chihuahua, Chih. Mexico.

D^az-Plascencia, D. 2017 (a). Technological innovation of an apple yeast additive. ED. Academica espanola.

D^az-Plascencia, D. 2017 (b). Manzarin in animal feed. ED. Academica Espanola.

Dong, G. F., Y. O. Yang, X. M. Song, L. Yu, T. T. Zhao, G. L. Huang, Z. J. Hu, and J. L. Zhang. 2013. Comparative effects of dietary supplementation with maggot meal and soybean meal in gibel carp (*Carassius auratus* gibelio) and darkbarbel catfish (*Pelteobagrus vachelli*): growth performance and antioxidant responses. Aquac. Nutr. doi: 10.1111/anu.12006

FAO. 2020. Proceedings of the International Symposium on Fisheries Sustainability: strengthening the science poly nexus. FAO headquarters, 18-21 November 2019, Rome, Italy. Fisheries and Aquaculture Proceedings No. 65. Rome. 116 pp.(also available at http://www.fao.org/3/ca9165 in pdf.

Gaylord, T. G., F. T. Barrows, S. D. Rawles, K. Liu, P. Bregitzer, A. Hang, D. E. Obert and C. Morris. Morris. 2009. Apparent digestibility of nutrients and energy in extruded diets from cultivars of barley and wheat selected for nutritional quality in rainbow trout. *Oncorhynchus mykiss.* Aquac. Nutr. 15:306-312.

Graminha, E. B. N., A. Z. L. Gonc and R. D. P. B. Pirota. 2008. Enzyme production by solidstate fermentation : Application to Animal Nutrition. 144:1-22.

Hoyos, Jose L, Villada, Hector S, Fernandez, Alejandro, & Ortega-Toro, Rodrigo (2017). Quality Parameters and Methodologies for Determining the Physical Properties of Extruded Fish Feeds. Informacion tecnologica, 28(5), 101-114. https://dx.doi.org/10.4067/S0718-07642017000500012.

Idenyi JN, Eya JC, Nwankwegu AS and Nwoba EG 2022. Aquaculture sustainability through alternative dietary ingredients: Microalgal value-added products. Engineering Microbiology: 100049. https://doi.org/10.1016/J. ENGMIC.2022.100049.

Joshi, V. K. and D. K. Sandhu. 1996. Preparation and evaluation of an animal feed byproduct produced by solid-state fermentation of apple pomace. Bioresour. Technol. 56:251255.

Kesarcodiwatson, A. and H. Kaspar. 2008. Probiotics in aquaculture: The need, principles and mechanisms of action and screening processes. Aquaculture 274:1-14.

Lambert, R. J. W., P. N. Skandamis, P. J. J. Coote and G. J. E. Nychas. 2001. A study of the minimum inhibitory concentration and mode of action of oregano essential oil, thymol and carvacrol. J. Appl. Microbiol. 91:453-462.

Leung, S., S. Barrington, Y. Wan, X. Zhao and B. El-Husseini. 2007. Zeolite (clinoptilolite) as feed additive to reduce manure mineral content. Bioresour. Technol. 98:3309-3316.

Li, Z. 2003. Use of surfactant-modified zeolite as fertilizer carriers to control nitrate release. Microporous Mesoporous Mater. 61:181-188.

Liu, X. Y., Y. Wang and W. X. X. Ji. 2011. Growth, feed utilization and body composition of Asian catfish (*Pangasius hypophthalmus*) fed at different dietary protein and lipid levels. Aquac. Nutr. 17:578-584.

Lovell, R. T. 1991. Nutrition of aquaculture species . R T Lovell. J. Anim. Sci. 69:4193-4200.

Navarrete, J. L. B., B. C. L. Jimenez and B. E. Bautista. E. Bautista. 2005. Seasonal rooting of oregano (*Lippia berlandieri* Schawer). Rev. Chapingo Ser. Zo. Aridas 4:25-30.

Pandey, A. 2003. Solid-state fermentation. Biochem. Eng. J. 13:81-84.

Renteria, P. M., R. R. Herrera, C. N. Aguilar and G. V. Nevarez-Moorillon. 2014. Microbiological effect of fermented Mexican oregano (*Lippia berlandieri* Schauer) waste. Waste and Biomass Valorization 5:57-63.

SAGARPA and CONAPESCA. 2012. Anuario estadistico de acuacultura y pesca 2011.

SAGARPA. 2009. Agro-economic Monitor 2009 of the state of Chihuahua.

Scalbert, A. and C. Manach. 2005. Dietary polyphenols and the prevention of diseases. Crit. Rev. Food Sci. Nutr. 45:287-306.

Secretary of Health. 1995. Norma oficial mexicana NOM-092-SSA1-1994 Goods and services. Method for counting aerobic bacteria on a plate.

Shaeffer, T. W., M. L. Brown, and K. A. Rosentrater. 2009. A. Rosentrater. 2009. Performance characteristics of Nile Tilapia fed diets containing graded levels of fuel based distillers dried graing with solubles. J. Aquac. Feed Sci. Nutr. 4:78-83.

Silva-Vazquez, R. 2015. Essential oil of Mexican oregano (*Lipia berlandieri* Schauer) in broiler chicken feed. PhD Thesis. Faculty of Zootechnics and Ecology. Autonomous University of Chihuahua. Chihuahua, Chih. Mexico.

Teixeira, B., A. Marques, C. Ramos, C. Serrano, O. Matos, N. R. Neng, J. M. F. Nogueira, J. A. Saraiva and M. L. Nunes. 2013. Chemical composition and bioactivity of different oregano (*Origanum vulgare*) extracts and essential oil. J. Sci. Food Agric.

Thiessen, D. L., G. L. Campbell and P. D. Adelizi. 2003. Digestibility and growth performance of juvenile rainbow trout (*Oncorhynchus mykiss*) fed with pea and canola products. Aquac. Nutr. 9:67-75.

Wang, Y., J. Li and J. Lin. 2008. Probiotics in aquaculture: Challenges and outlook. Aquaculture 281:1-4.

Wang, Y.-B. 2007. Effect of probiotics on growth performance and digestive enzyme activity of the shrimp *Penaeus vannamei*. Aquaculture 269:259-264.

Yamashita, Y., T. Katagiri, N. Pirarat, K. Futami, M. Endo and M. Maita. 2009. The synthetic antioxidant, ethoxyquin, adversely affects immunity in tilapia (*Oreochromis niloticus*). Aquac. Nutr. 15:144-151.

USE OF TWO NATURAL ANTIMICROBIAL ADDITIVES IN THE SOLID-STATE FERMENTATION OF APPLE BAGASSE

SUMMARY

USE OF TWO NATURAL ANTIMICROBIAL ADDITIVES IN THE
SOLID-STATE FERMENTATION
OF APPLE BAGASSE

The antimicrobial effect of oregano essential oil and zeolite on solid-state fermentation (SSF) of apple bagasse was evaluated. A 5 % clinoptilolite zeolite (ZEO), 0.1 % oregano essential oil (AEO) and both (ZXA) were used against the traditional method of molasses broth added with urea, ammonium sulphate and minerals as control (CTL). There were three replicates per treatment in 1-litre tanks, with sampling at 0, 6, 12, 24, 48, 72 and 96 h. The variables evaluated were microbial count, microbial growth rate, microbial growth rate and microbial growth rate. The variables evaluated were microbial plate count, pH and yeast count with Neubauer chamber. The data were analysed with a repeated means over time model and multiple means comparison. Maximum yeast growth was obtained at 48 h in ZXA (462×10^6 cells/g), while in CTL at 96 h (470.5×10^6 cells/g), reducing fermentation time and optimising the process. Aerobic bacterial counts were lower ($p<0.01$) at 48 h, from $0.70 \times 10 \times 10^6$ CFU/g in CTL to 0.25×10^6 CFU/g in ZXA, with AEO having the lowest bacterial count (0.11×10^6 CFU/g). The pH was higher ($p<0.01$) in ZEO (4.80) and ZXA (4.62), versus CTL (4.06) and AEO (4.03). It is concluded that the use of both additives provides microbiological advantages over the traditional FES process, decreasing fermentation time and avoiding bacterial contamination, guaranteeing a process free of microbiological contaminants.

INTRODUCTION

Humanity is currently faced with the challenge of producing enough food to sustain the enormous population growth that has occurred in recent years. However, this has led to more intensive harvesting, the search for more efficient animal breeds and, above all, the production of more waste when producing food for the population. The state of Chihuahua is the leading national apple producer, with 43.1 % of the national planted area and 73.7 % of the total production, producing 370,040.20 tonnes per year (SAGARPA, 2014), with which it can be estimated that a quarter of this is waste apple that does not meet quality standards. This waste, together with the by-products generated by the apple processing industry, generates waste, mainly apple bagasse, accounting for 25 to 30 % of the weight of the fresh fruit, which represents a serious pollution problem, as it is highly biodegradable and significantly alters the environment in which it is discarded thanks to its high content of carbohydrates, acids, fibres, vitamin C and minerals (Joshi and Sandhu, 1996).

Thanks to the low nitrogen content of apple bagasse, its inclusion in animal diets has been difficult, a situation that makes it imperative to develop techniques that allow its consumption and reduce the environmental impact it generates. Therefore, one of the most successful alternatives has been the solid-state fermentation (SSF) of bagasse, added with urea nitrogen and mixtures of vitamins and minerals that allow the development of yeasts, which in addition to degrading the fibre, can provide protein of microbial origin, i.e. supplement at low cost high quality protein (Dhillon *et al.*, 2013).

In solid state fermentation, physicochemical parameters must be controlled to avoid contamination, as apple bagasse provides a medium with high moisture and a large amount of available carbohydrates (Castillo *et al.*, 2011).

This raises the option of using by-products through FES to maintain a suitable microbiological environment, reduce microbiological contamination, but without disturbing the development of fermentative yeasts.

Oregano essential oil (OEO) has been used as a natural alternative to artificial preservatives, given its antioxidant activity, safety, antioxidant capacity and, above all, its antimicrobial capacity (Teixeira *et al.,* 2013). Thanks to these properties, it is considered that the addition of AEO to apple bagasse FES will provide a defence against contamination by bacteria that will consume the substrates and decrease the fermentative efficiency of the yeasts.

Zeolites are clayey materials obtained from sedimentary deposits. Of the more than 40 types of naturally occurring zeolites, clinoptilolite is particularly functional in adsorbing nitrogen compounds, especially NH_4^+. Clinoptilolite is a type of zeolite that has an open molecular structure, with a pore volume of about 35 % and is easily found in mineral deposits worldwide. This characteristic is obtained thanks to the three-dimensional structure of aluminosilicates with four- and five-membered rings, thus forming micropores capable of harbouring exchangeable cations such as NH_4^+, Na^+, K^+ and Ca^{2+} (Leung *et al.*, 2007).

Therefore, the aim of this work was to evaluate the microbiological effect of the addition of micronised zeolite and oregano essential oil to the solid state fermentation of apple bagasse, as well as to determine the adsorptive effect of zeolite on nitrogen compounds in apple.

MATERIALS AND METHODS

The experiment was carried out in a greenhouse of the Faculty of Zootechnics and Ecology of the Autonomous University of Chihuahua, where apple bagasse was kept in plastic containers of one litre capacity, to which urea and ammonium sulphate were added as a source of nitrogen and a mineral supplement to provide the nutrients necessary for yeast growth (control treatment).

The raw material used was apple bagasse from CONFRUTTA, a company dedicated to the production of apple juice from Ciudad Cuauhtemoc, Chihuahua, using Golden Delicious apples.

The control treatment consisted of apple bagasse without additives, while the AEO treatment was supplemented with 0.1% oregano essential oil (AEO), 0.1% oregano essential oil (AEO) , 0.5% oregano essential oil (AEO) and 0.5% oregano essential oil (AEO).

The ZEO treatment was supplemented with 5 % micronised zeolite and finally, the ZXA treatment included both oregano essential oil and zeolite, with three replicates per treatment, as shown in Table 2.

The zeolite used was obtained commercially, from a mine in San Luis Potos^ Mexico, with a porosity of 45 to 52 %, a density of 700 to 850 kg/m^3 , a hardness of 2 to 3 Mohs, a water absorption capacity of 42 to 50 %, pores of 4 to 7 A and a melting point of 1300 °C. The supplier's guaranteed composition of the zeolite is given in Table 3.

Once the mixture was made, it was homogenised and four manual agitations were made per day, stirring the product completely to generate adequate oxygenation for four days. Samples were taken at 0, 6, 12, 12, 24, 48, 72 and 96 h for yeast counts, and at 0, 48 and 96 h for bacterial counts.

Table 2. Formulations of the apple bagasse FES treatments

Treatment[1]	Urea %	Sulfate of ammonium %	Mineral supplement %	Zeolite %	AEO %
	Urea	ammonium %	supplement %	Zeolite	AEO %
		ammonium %	Mineral		AEO %
		ammonium %	supplement %		AEO %
		ammonium %	Mineral		AEO %
		ammonium	supplement %		AEO %
			Mineral		AEO
			supplement %		
			Mineral		
			supplement %		
			Mineral		
			supplement %		
			Mineral		
			supplement %		
			Mineral		
			supplement %		
			Mineral		
			supplement %		

			Mineral supplement		
1.- Control	2.0	0.4	1.0	-	-
2.- ZEO	2.0	0.4	1.0	5.0	-
3.- AEO	2.0	0.4	1.0	-	0.1
4.- ZXA	2.0	0.4	1.0	5.0	0.1

[1] ZEO indicates treatment with 5 % Zeolite, AEO treatment with 0.1 % Oregano Essential Oil and ZXA treatment with both (5 and 0.1 % respectively).

Table 3.- Chemical composition of clinoptilolite zeolite[1]

Compound	Concentration % Concentration % Concentration % Concentration
Silicon oxide (SiO_2)	67.9
Aluminium oxide (Al_2O_3)	13.7
Potassium oxide (K_2O)	4.7
Sodium oxide (Na_2O)	2.2
Calcium oxide (CaO)	1.7
Ferric oxide (Fe_2O_3)	1.8
Magnesium oxide (MgO)	1.8

[1] Analysis provided by the chemical analysis laboratory of the metallurgy institute of the U.A.S.L.P.

Variables Assessed

Bacterial count. The microbiological count was carried out in accordance with the Official Mexican Standard NOM-092-SSA1-1994 Method for counting aerobic bacteria on a plate. Five samples were taken from different places in the FES, and homogenised, one gram was taken, diluted 6 times in diluent medium, and inoculated in each Petri dish, running all samples in triplicate. After inoculating the dilutions of the samples into the Petri dishes, add 12 to 15 mL of the prepared standard bead medium, mix by 6 movements from right to left, 6 clockwise, 6 counterclockwise and 6 back and forth, on a smooth, horizontal surface until the inoculum is completely incorporated into the medium and allowed to solidify. One box without inoculum is included for each medium and prepared diluent as a sterility control. The boxes are incubated in an inverted position at 35 ± 2 °C for 48 ± 2 hours.

In the readout, plates with 25 to 250 CFU are selected and all the colonies developed on the selected plates are counted. After incubation, the plates in the range of 25 to 250 colonies are counted using the colony counter and recorder. The average count per gram of the dilution is calculated.

Yeasts. The yeast count was carried out by a simple count in an improved Neubauer chamber, by the method described by D^az-Plascencia (Diaz-Plascencia, 2011), where a sample of 1 g of the product was taken, which was taken to serial dilutions in phosphate buffer solution, and then placed in a Neubauer chamber, for the counting of yeasts (individual or in gelation) in the four quadrants of the streaking. From this data, the amount of yeast per gram of apple bagasse ferment is calculated.

pH. The pH was measured throughout the experiment with a Hanna digital potentiometer, which was sterilised by immersion in alcohol, dried and immersed in the broth to take the measurement according to the manufacturer's instructions.

Antioxidant activity. To determine the antioxidant activity of fermented apple bagasse, the DPPH (1, 1-diphenyl-2-picrylhydrazalium) method was used according to Ajila, (Ajila *et al.*, 2011), for which he mixed the fermented apple bagasse extract (200 p.l) with 1 mL of the DPPH solution. After being homogenised in vortex, it was left to react for 20 minutes in total darkness. The absorbance was read at 517 nm, with which the free radical scavenging activity in percentage was calculated with the equation (% scavenging activity) = 1 - (A_s /Ao) x 100, where Ao is the absorbance of the control while As is the absorbance of the sample.

Protein. The protein variable was determined by the methodology of Fagbenro, who used the trichloroacetic acid precipitation method to determine true protein and discern it from non-protein nitrogen (Fagbenro and Jauncey, 1998).

Statistical Analysis

The data obtained were analysed by means of a repeated means model over time, using the Mixed procedure of SAS 9.1.6 (SAS, 2006) and the comparison of means was carried out by means of Least Square Means, from the same software package.

RESULTS AND DISCUSSION

A significant effect (P < 0.01) of the inclusion of AEO and Zeolite clinoptilolite was found for the variable of mesophilic aerobic bacteria in the FES of apple bagasse, decreasing the bacterial count with respect to the control at hour 48 and creating a significant difference between the treatments that contained AEO with respect to those that did not (P < 0.01) at hour 96. These results can be explained by the antimicrobial effect of AEO, observing the decrease in the total number of bacteria thanks to the antibacterial activity of the thymol and carvacrol contained in AEO, which is in agreement with the work of Lambert *et al.* (2001), who inhibited bacterial growth with concentrations of 5% thymol in AEO.

There was also an interaction between treatments and time (P < 0.01), so it can be said that as bacteria develop in the apple bagasse FES, they are affected by the presence of AEO or ZEO, so that their numbers are reduced compared to the control treatment.

At 48 h of fermentation, the difference between the Control and ZEO treatments, with respect to the other two containing AEO, is clearly observed (Graph 1), which indicates that, at the moment of maximum bacterial growth, these are significantly inhibited by AEO (P < 0.01). This effect of AEO is observed throughout the fermentation, making clear the inhibitory effect of AEO on aerobic mesophilic bacteria.

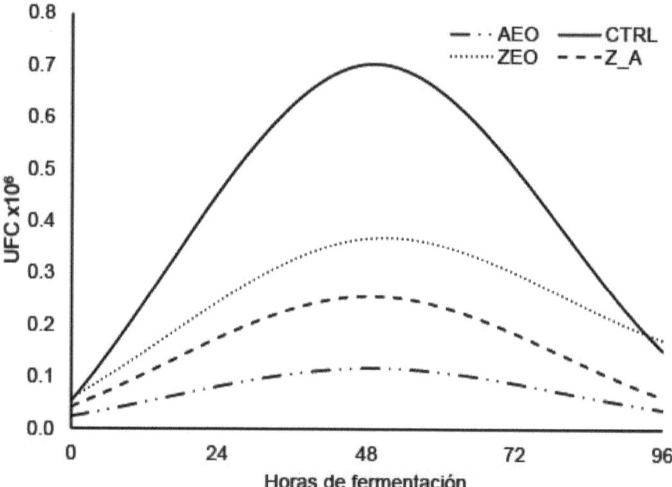

Means of least squares of the total mesophilic aerobic bacteria count (CFU $\times 10^6$) in the FES of apple bagasse added with AEO and ZEO.

The quantified yeasts presented the characteristic multiplication of these microorganisms, consistent with the growths presented by other authors (DiazPlascencia, *et al.*, 2012) and it is observed at time 0 of the FES that there is a significant effect (P < 0.01) by the addition of the treatments, observing a significant decrease in the amount of yeasts (Table 4), so it is clear that both Zeolite and AEO have an effect on the yeast count.

In the first 24 h a poor yeast growth can be observed in the ZXA treatment (P < 0.01) with respect to the other treatments, recovering at h 48, where this treatment presents the highest yeast count, comparable only to the final count of the control treatment. This may indicate an adequate utilisation of the fast substrates present in the apple bagasse and a subsequent exhaustion of these substrates, since from this inflection point onwards, the yeast count of the ZXA treatment decreases sharply. The three treatments present their maximum peak of yeast growth at 48 h, and from then on there is a decrease of the count, while the control presents the highest count at 96 h.

At the end of the experiment, a significant difference can be found between the control treatment and the others, with the control having a higher count, followed by the ZEO treatment, which from h 48 stabilised its growth and maintained its count relatively stable. However, the two treatments containing AEO had a significant decrease in their yeast population, evidencing the antimicrobial power of AEO (Lambert *et al.*, 2001).

Table 4.- Least squares means (±EE) of the total yeast count (Yeast $\times 10^6$ /g) in the FES of apple bagasse added with AEO and ZEO. [1,2]

Fermentation time, h	Treatment[1 2 3]				±EE
	Control	ZEO	AEO	ZXA	
0	40.00^a	18.50^b	25.50^b	27.00^b	3.04
6	94.00^a	37.50^b	37.00^b	34.00^b	4.26
12	$111.50a$	96.00^b	106.50^{ab}	45.00^c	2.76
24	$152.00a$	$143.50a$	$159.50a$	74.50^b	7.53
48	362.00^b	$339.5.00^b$	384.50^b	$462.00a$	14.83
72	$447.00a$	337.00^b	247.50^c	273.00^c	14.22
96	$470.50a$	326.50^b	281.00^{bc}	247.50^c	17.25

The measurement of pH during the experiment shows that there is an effect of zeolite in maintaining a higher pH, as the two treatments that include zeolite have a significantly higher pH than the other treatments (P < 0.01). The effect of zeolite can be clearly appreciated, as it maintains a stable pH from the moment it is added to the FES, since its pH regulating activity is present from h 0 of fermentation, statistically separating the ZEO and ZXA treatments from those that do not contain zeolite in their formulation (P < 0.01). Graph 2 shows the pH fluctuation that occurred in the AEO and Control treatments during the first 24 h, so that the effect of ammonium adsorption on the pH of the zeolite is evident. These results are similar to those

---

[1] Different literals between columns indicate significant difference (P < 0.01).
[2] Literals differ between rows.
[3] ZEO indicates treatment with 5 % Zeolite, AEO treatment with 0.1 % Oregano Essential Oil and ZXA treatment with both (5 and 0.1 % respectively).

presented by other authors, as they explain the zeolite's capacity to adsorb ammonium molecules and other ions, thus regulating the pH for the benefit of the microbiological system (Forte and Maugeri, 2007; Kardaya *et al.*, 2012).

The antioxidant activity of fermented apple bagasse increased as the fermentation time elapsed (Figure 3), doubling its activity from 6.44 to 12.43 %, which is consistent with that presented by Ajila *et al.* (2011), as the increase in antioxidant capacity can be demonstrated by two main effects: as fermentation is carried out in the solid state, the substrate is stirred to aerate, which breaks the stems, seeds and apple cells, releasing antioxidant compounds; on the other hand, the same yeast growth causes the enzymes of these to digest cell walls, membranes and release compounds with antioxidant activity from the apple tissues.

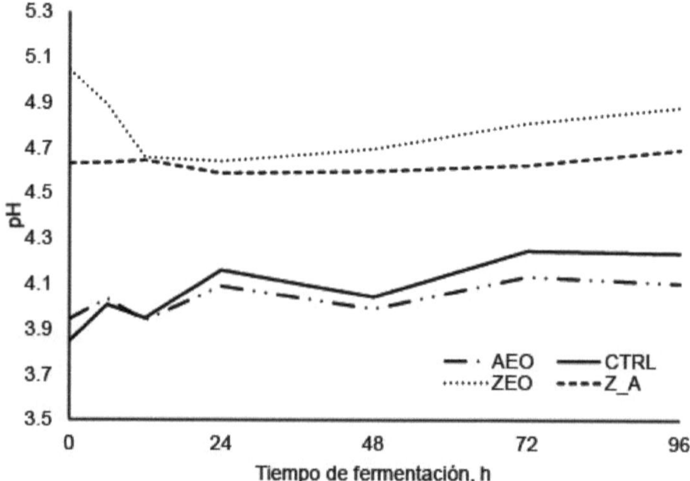

Means of least squares of pH of the FES of apple bagasse added with AEO and ZEO.

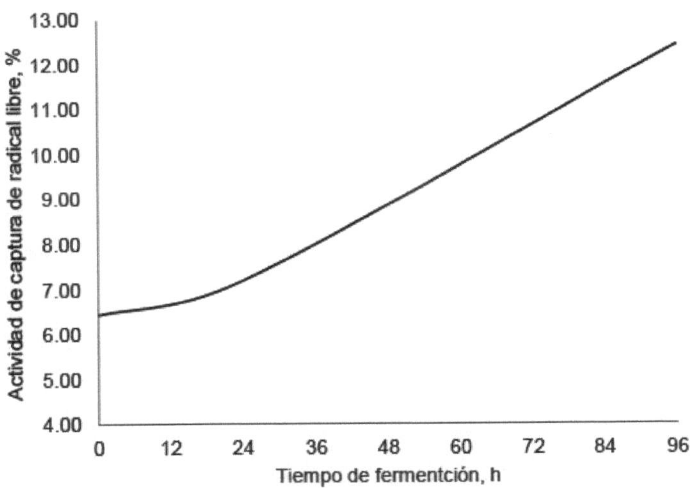

Free radical capture activity in percentage of fermented apple bagasse.

The results of true protein show that there was an increase of true protein during the course of fermentation, almost doubling the content of true protein in the final product. In spite of having a similar behaviour, in the last stage of fermentation it was possible to differentiate perfectly between treatments ($P < 0.05$) at 72 and 96 hours (Table 5), differentiating perfectly between the control and the other treatments. The maximum protein concentration in fermentation occurred at hour 48 for all treatments, except the control, where it was present until hour 96. This work is in agreement with the work of Bhalla and Joshi (1994), who using controlled strains of *Saccharomyces* achieved protein concentrations of up to 30 %, but the results were lower than those of Pirmohammadi *et al.* (2006), who managed to obtain protein concentrations of up to 40.1 %, although in this work no additives from other carbon sources were used, as in the aforementioned work.

From the growth curves generated from the bacterial and yeast counts, it can be concluded that at 48 h the rapidly metabolised substrates such as simple carbohydrates are exhausted, as a marked decrease in the growth rate is observed, as well as in the number of yeasts and bacteria, so it is understood that the microorganisms are forced to use other substrates as a source of energy for their development.

Table 5.- True protein over time of the FES of apple bagasse added with AEO and ZEO [1,2]

Fermentation time (h)	Treatment[4 5 6]				±EE
	AEO	AXZ	Control	ZEO	
0	18.51^{ab}	18.74^a	18.10^b	18.60^{ab}	0.1333
6	18.73^a	19.10^a	18.72^a	19.09^a	0.1219
12	20.15^a	19.32^b	19.20^b	20.15^a	0.1778
24	22.65^a	19.94^b	23.13^a	22.57^a	0.1851
48	30.49^a	30.81^a	30.47^a	30.25^a	0.2822
72	29.21^c	28.66^c	33.83^a	30.26^b	0.1466
96	29.23^c	28.78^c	34.42^a	30.74^b	0.2445

In the control treatment this inflection point was not observed, so when observing the slope of the control treatment, it is seen that the growth was slower, so that the inflection point, which is when the fast metabolising substrates were exhausted, occurs up to 72 h, so it is established that it is less efficient, as at 96 h it has a yeast count of 470.5 x10^6 /g, equivalent to that of the ZXA treatment at 48 h (462 x10^6 /g), so with the addition of Zeolite and Essential Oil of Oregano the result is obtained in less time, improving the process.

It was also observed that the appearance of the treatments that included zeolite was much drier and earthier than the others, thanks to the chemical characteristics of zeolite, as it had a drying effect, adsorbing water and limiting the growth of microorganisms by decreasing the water activity in the FES, a situation that easily affects bacteria, but yeasts manage to resist better. This effect can be seen in the lower development in the first h, as the zeolite treatments had a slower microbiological development compared to the other treatments.

[4] Different literals between columns indicate significant difference (P < 0.01).
[5] Literals differ between rows.
[6] ZEO indicates treatment with 5 % Zeolite, AEO treatment with 0.1 % Oregano Essential Oil and ZXA treatment with both (5 and 0.1 % respectively).

CONCLUSIONS AND RECOMMENDATIONS

It is concluded that adding solid-state fermentation of apple bagasse with zeolite and oregano essential oil provides productive advantages over the control treatment, as maximum yeast development is achieved in a shorter time by adding both AEO and clinoptilolite zeolite, compared to the control treatment, which achieves a faster, safer and more stable process.

It is concluded that the addition of oregano essential oil causes a decrease in the aerobic mesophilic bacteria count in the solid state fermentation, preventing bacterial contamination.

The addition of oregano essential oil and zeolite did not affect the production of true protein, so the efficiency of the solid state fermentation was not reduced. It is also concluded that the pH values remained constant in the treatments containing zeolite, so zeolite has a pH regulating effect as a buffer.

LITERATURE CITED

Ajila, C. M., F. Gassara, S. K. Brar, M. Verma, R. D. Tyagi and J. R. Valero. 2011. Polyphenolic antioxidant mobilization in apple pomace by different methods of solid-State fermentation and evaluation of Its antioxidant activity. Food Bioprocess Technol. 5:2697-2707.

Bhalla, T. C. and M. Joshi. 1994. Protein enrichment of apple pomace by co-culture of cellulolytic moulds and yeasts. World J. Microbiol. Biotechnol. 10:116-7.

Castillo, Y., O. Ruiz, C. Angulo, C. Rodriguez, A. EKas and O. La O. 2011. Inclusion of bakery residues in some metabolites and bromatological indicators of solid-state fermentation of apple bagasse. Cuban Journal of Agricultural Science. 45 141-144.

D^az-Plascencia, C. Rodriguez-Muela and P. Mancillas-Flores. 2012. In vitro fermentation of fodder cactus with Kluyveromyces lactis yeast inoculum obtained from waste apple. Rev. Electron. Vet. 13:1-11.

D^az, D., F. Salvador, O. Ruiz, C. Arzola, A. Flores, O. La O and A. EKas. 2010. Effect of the level of urea and soybean paste on the concentration of proteins during solid-state fermentation of waste apple (*Malus domestica*). Revista Cubana de Ciecia Agricola 44 23-26.

D^az-Plascencia, D. 2011. Development of a yeast-based inoculum and its effect on in vitro fermentation kinetics in rations for high-producing Holstein cows. PhD thesis. Faculty of Zootechnics and Ecology. Autonomous University of Chihuahua. Chihuahua, Chih. Mexico.

Fagbenro, O. and K. Jauncey. 1998. Physical and nutritional properties of moist fermented fish silage pellets as a protein supplement for tilapia (*Oreochromis niloticus*). Anim. Feed Sci. Technol. 71:11-18.

Forte, M. and F. Maugeri. 2007. Purification of clavulanic acid from fermentation broth using zeolites. J. Biotechnol. 131:S191.

Joshi, V. K. and D. K. Sandhu. 1996. Preparation and evaluation of an animal feed byproduct produced by solid-state fermentation of apple pomace. Bioresour. Technol. 56:251 - 255.

Kardaya, D., D. Sudrajat and E. Dihansih. 2012. Dihansih. 2012. Efficacy of dietary urea-impregnated zeolite in Improving rumen fermentation characteristics of local lamb. Indones. J. Anim. Sci. Technol. 35:207-213.

Lambert, R. J. W., P. N. Skandamis, P. J. Coote and G. J. E. Nychas. 2001. A study of the minimum inhibitory concentration and mode of action of oregano essential oil, thymol and carvacrol. J. Appl. Microbiol. 91:453-462.

Leung, S., S. Barrington, Y. Wan, X. Zhao and B. El-Husseini. 2007. Zeolite (clinoptilolite) as feed additive to reduce manure mineral content. Bioresour. Technol. 98:3309-3316.

Pirmohammadi, R., Y. Rouzbehan, K. Rezayazdi and M. Zahedifar. 2006. Chemical composition, digestibility and in situ degradability of dried and ensiled apple pomace and maize silage. Small Rumin. Res. 66:150-155.

SAGARPA. 2009. Agro-economic Monitor 2009 of the state of Chihuahua.

SAS Institute, Inc (2006) SAS/STAT users guide: Statics Version 9 Cary, North Carolina. U.S.A.

Secretary of Health. 1995. Norma oficial mexicana NOM-092-SSA1-1994 Bienes y servicios. Method for counting aerobic bacteria on a plate.

Teixeira, B., A. Marques, C. Ramos, C. Serrano, O. Matos, N. R. Neng, J. M. F. Nogueira, J. A. Saraiva and M. L. Nunes. 2013. Chemical composition and bioactivity of different oregano (*Origanum vulgare*) extracts and essential oil. J. Sci. Food Agric.

- USE OF TWO NATURAL ANTIMICROBIAL ADDITIVES IN THE PRODUCTION OF A PROBIOTIC YEAST INOCULUM

SUMMARY

USE OF TWO NATURAL ANTIMICROBIAL ADDITIVES IN THE PRODUCTION OF A PROBIOTIC YEAST INOCULUM

The aim of the present work was to evaluate the antimicrobial effect of oregano essential oil (OEO) on the production of probiotic yeasts by the growth method in molasses broths, as well as to measure the pH buffering effect of zeolite. A system of plastic containers was prepared with the *Kluyveromyces lactis* inoculated molasses broths, which were aerated, at a temperature of 25 ± 0.5 °C, and measurements of pH, aerobic mesophilic bacteria and yeast counts per ml of broth were taken. It was determined that broths with AEO had a lower bacterial count (1.1×10^4 CFU/mL) than the control (1.6×10^4 CFU/mL), while treatments that included zeolite were significantly lower (1.0×10^4 CFU/mL). With regard to yeasts, they developed in the AEO treatment, obtaining their maximum concentration at 72 hours (3.99×10^7 yeast/mL), but remained below the control treatment (5.2×10^7 yeast/mL). The zeolite, due to its ability to capture nitrogen compounds, prevents bioavailability for microbial growth, affecting bacteria and yeasts, thus decreasing the yield of the broths. It is concluded that the inclusion of AEO adds antioxidant and antimicrobial benefits for the hosts. Thus, an inexpensive, yeast-rich, antioxidant-rich and safe probiotic is obtained.

INTRODUCTION

Liquid media for fermentation or submerged solid fermentation are systems that have been in use for many years, standing out for their versatility in terms of handling and product yield, in contrast to solid fermentation systems, where it is difficult to control all variables as easily as in broths (Krishna, 2005).

Yeasts are unicellular organisms found almost everywhere on the planet, and are naturally disseminated by air, water and food, so it is a natural occurrence for them to be found in the gastrointestinal tract of aquatic species. Studies concerning these microorganisms have found that they are beneficial to aquatic organisms, as they have components, such as p-glucans, which are used as immunostimulants, and more recently, their use as probiotic agents in fish diets has been studied (Gatesoupe, 2007; Abdel-Tawwab *et al.*, 2008).

Oregano essential oil (OEO) has been used as a natural alternative to artificial preservatives, given its antioxidant activity, safety, antioxidant capacity and, above all, its antimicrobial capacity. Thanks to these properties, it is considered that the addition of AEO to molasses and urea-based broths for yeast growth will provide a defence against contamination by bacteria that would consume the substrates and could decrease the productive efficiency of the yeasts (Teixeira *et al.*, 2013).

Zeolites are clay materials obtained from sedimentary deposits. Of the more than 40 types of naturally occurring zeolites, clinoptilolite is particularly functional as it has the ability to adsorb nitrogenous compounds, especially NH_4^+ (Li, 2003). Clinoptilolite is a type of zeolite that has an open molecular structure, with a pore volume of about 35 % and is easily found in mineral deposits worldwide. This characteristic is obtained thanks to the three-dimensional structure of aluminosilicates with four- and five-membered rings, forming micropores capable of hosting exchangeable cations such as NH_4^+, Na^+, K^+ and Ca^{2+} (Leung *et al.*, 2007). This characteristic of zeolite can improve the

nutrient utilisation during submerged solid fermentation and improved yeast growth.

Therefore, the main objective of this study is to evaluate the antimicrobial effect of Oregano Essential Oil on the traditional production of probiotic yeasts in a molasses broth added with nitrogen and sulphur. As particular objectives, the aim is to measure the pH buffering effect provided by the clinoptilolite zeolite and to determine the time of maximum yeast count as the end point of the yeast production process in the product.

MATERIALS AND METHODS

The present experiment was carried out in the microbiology laboratory of the Facultad de Zootecnia y Ecolog^a of the Universidad Autonoma de Chihuahua, for which culture broths based on molasses were prepared with the nutrients necessary for the reproduction of *Kluyveromyces lactis* yeasts in three treatments and a control. The control treatments (CTL), which is just the traditional nutrient broth of molasses, nitrogen compounds and sulphur, were used (Table 6). The ZEO treatment, CTL plus 2 % of the total weight in clinoptilolite zeolite; the AEO treatment, CTL plus 0.1 % of the total weight in Oregano Essential Oil (AEO); and the AXZ treatment, consisting of CTL plus a combination of both additives (0.1 % AEO and 2 % zeolite).

The broths were placed in one-litre plastic containers, with an aeration stone connected to a constantly running air pump, in order to ensure adequate oxygenation and a current generated by the bubbling, so as to have complete homogenisation of the broth. These vessels were placed inside a Thermo Fischer Scientific "Precision 815" incubator, which maintained a constant temperature of 25 ± 0.5 °C throughout the experiment.

The yeasts used were the *Kluyveromyces lactis* strain, retrieved from the faculty's microorganism bank. To reactivate the yeasts, a nutrient broth of meat extract, soy peptone, malt extract and necessary additives (Table 7), prepared specifically for this purpose, was used (Table 6). Composition of the molasses nutrient broths used in the experiment.

			Treatment[VII]		
Ingredient	**Control, %AEO**	**, %ZEO**	**, %AXZ**	**, %**	
Molasses		24.024	.024.024	.	0
Urea	1. 01.	01.01. 0			
Ammonium sulphate	0. 20.	20.20. 2			
Yeast inoculum	3. 03.	03.03. 0			
AEO	-0.1-0	.1			
Zeolite	--2.	02.0			

Table 7. Chemical composition of the yeast reactivation broth

Compound	Content per litre of stock[1]
Meat extract	10.0 g
Soy peptone	10.0 g
Sodium chloride	5.0 g
Dextrose	5.0 g
Malt extract	3.0 g
Soluble starch	1.0 g
L-Cisteha	0.5 g

[1] Dissolve 34.5 g in 1 l of water and heat with stirring until dissolved. Sterilise at 121 °C for 15 minutes.

[VII] AEO indicates treatment with 0.1 % Oregano Essential Oil, ZEO treatment with 2 % Zeolite and AXZ treatment with both (0.1 and 2 % respectively).

The yeasts were maintained for 96 h in this broth at 25 °C and a production of 5.7 x 10^7 cells/mL was achieved.

The molasses nutrient broth was then inoculated with these yeasts at a rate of 0. 1% of the total weight. This inoculated broth was then left to air and cool to room temperature.

controlled for 72 h in the incubator, in order to maintain a stable environment.

Response Variables

The variables evaluated were total aerobic bacterial plate count (CFU/mL broth), amount of sugar in the broth (°Bx), pH and yeast count (Cells/mL).

Yeast count. The yeast count was carried out by a simple count in an improved Neubauer chamber, by the method described by D^az-Plascencia (2011), where a sample of 1 mL of broth was taken, which was taken to serial dilutions, and then placed in a Neubauer chamber, where the yeasts were counted (individual or in gems) in the four quadrants of the streaking. From this data, the amount of yeast per mL of molasses nutrient broth was calculated.

Bacteriological count. The bacteriological count was carried out in accordance with the Official Mexican Standard NOM-092-SSA1-1994 Method for counting aerobic bacteria on a plate. One mL of the nutrient broth sample was taken, diluted 6 times in the diluent medium, and inoculated in each Petri dish, running all samples in triplicate. After inoculating the dilutions of the samples into the Petri dishes, add 12 to 15 mL of the prepared standard medium, mix by 6 movements from right to left, 6 clockwise, 6 anticlockwise and 6 back and forth, on a smooth, horizontal surface until complete incorporation of the inoculum into the medium is achieved and allowed to solidify. One box without inoculum is included for each medium and prepared diluent as a sterility control. The boxes are incubated in an inverted position at 35 ± 2 °C for 48 ± 2 hours.

After incubation, plates in the range of 25 to 250 colonies are counted using the colony counter and recorder. The average count per mL of the dilution is calculated.

pH. The pH measurements were taken with a Hanna digital potentiometer, immersing it in the broth and shaking it while waiting for it to reach the stability point and mark the reading on the display.

Soluble solids. The measurement of sugar concentration was carried out with a Hanna digital refractometer, to which a drop of broth is added to the reading glass, the measurement is activated and automatically gives the reading in Brix degrees.

Experimental Design

Four treatments were established: the treatment with oregano essential oil (AEO), the treatment with zeolite (ZEO) and the interaction of these treatments (AXZ), in addition to the control, which was the traditional method (CTL).

A completely randomised design was used, with four replicates for each treatment. Three samples were obtained from each replicate at 0, 6, 24, 48 and 72 h, giving 12 samples per treatment at each measurement time. **Statistical Technique for Hypothesis Testing**

A one-way ANOVA test was used to compare the variables of aerobic bacterial count, pH, sugar content and yeast count in the broth, in addition to a Tukey's mean comparison between treatments, considering the treatments statistically significant at $P < 0.05$. For the growth of yeasts and bacteria with respect to time, a study of repeated measurements over time was carried out with the Proc Mixed procedure, finding significant effects at $P < 0.05$. The statistical treatment was carried out with the computer package SAS, version 9.1.3 (SAS, 2006).

RESULTS AND DISCUSSION

The behaviour of the yeast count per mL of molasses broth developed in a linear way, increasing as time progressed, although presenting a greater increase of microorganisms per unit of time in the CTL (P > 0.05), which represents the traditional method of developing yeasts in molasses broth, where a maximum count of 5.27x107 yeast cells per mL was reached, followed by the AE.05), which represents the traditional method of developing yeasts in molasses broth, where a maximum count of 5.27x107 yeast cells per mL was reached, followed by the AEO, which had a longer adaptation phase, because until 48 h the exponential growth began (Table 8) and had a maximum count of $3.99x10^7$ cells/mL of broth. On the other hand, the AXZ and ZEO treatments, added with Zeolite clinoptilolite had a markedly slow yeast development (P > 0.05), and obtained a maximum yeast count of only $1.8x10^7$ and $1.9x10^7$ cell/mL respectively. This yeast development can be reflected in the concentration of soluble solids, as the initial values, close to 21 °Bx, decreased considerably (Figure 4), which means that the yeasts made use of the available sugars for their multiplication. However, the main decreases occurred in the CTL and AEO treatments, which is consistent with these treatments being those with the highest total yeast counts. While the AXZ and ZEO treatments maintained the most stable soluble solids concentration, which is also consistent with the lower yeast growth.

Table 8. Yeast concentration per mL of broth, expressed as yeast x 106/mL broth per h

h	AEO	AXZ	CTL	ZEO	E.E.
			Treatment[1]		
6	15.63^b	1.35^c	20.75^a	1.10^c	1.6404
24	19.25^b	4.05^c	33.75^a	6.20^c	1.3941
48	31.78^b	6.40^c	41.65^a	8.65^c	1.7383
72	39.96^b	10.85^d	52.70^a	19.15^c	1.7058

[1] AEO indicates treatment with 0.1 % Oregano Essential Oil, ZEO treatment with 2 % Zeolite and AXZ treatment with both (0.1 and 2 % respectively). CTL indicates Control treatment.

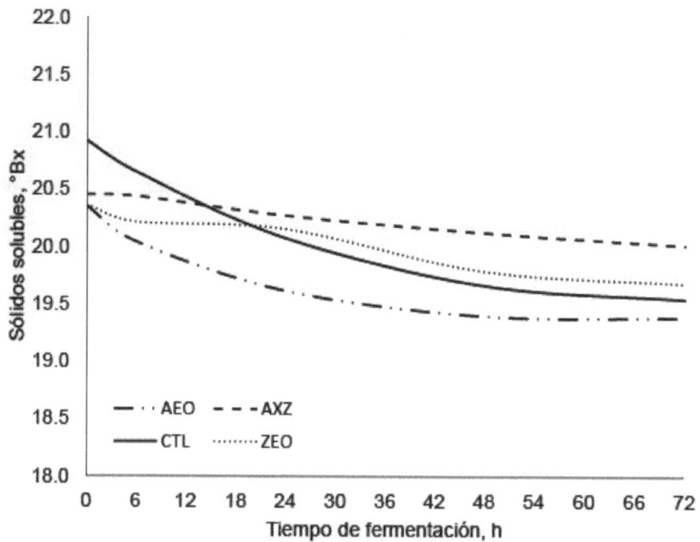

Soluble sugars in the broth versus time of the experiment, expressed in °Bx.

The concentration of hydrogen ions, measured as pH, remained relatively constant, as there was no great fluctuation throughout the treatments, except for the initial modification that occurred with the addition of oregano essential oil to the AEO and AXZ treatments, which brought the pH values to numbers close to 5 (Graph 5), while the CTL and ZEO treatments, which did not have AEO, remained at very stable values at a pH of 4 (P > 0.05). These results are consistent with the low bacterial growth, as adequate oxygenation and proper movement and homogenisation prevented stagnation of the broth, which would have generated zones of anaerobiosis, allowing anaerobic fermentation to lactic acid and alcohol, factors that would have altered the acidity of the broth (Grewal and Kalra, 1995).

In the case of the presence of bacteria, these showed little development, as they did not exceed a concentration of 2.11×10^4 CFU/mL, which is low compared to the work of other authors (Chun *et al.*, 2005), as in this experiment the conditions of agitation and constant oxygenation were maintained, in addition to a pH close to 4, so there were few conditions for the development of total mesophilic anaerobic bacteria. In the CTL treatment there was more bacterial growth (P > 0.05), with 2.11x104 CFU/mL, as it did not contain any additive that inhibited their growth, except for the physicochemical parameters conferred by the molasses and the same competition for nutrients against the yeasts. In second place, the AEO treatment was presented, which had a maximum concentration of 1.7×10^4 CFU/mL, being below the control, while, in a very similar situation, the AXZ and ZEO treatments had a lower bacterial development (P > 0.05), not exceeding 1.08×10^4 CFU/mL (Table 9), which indicates that the addition of zeolite hinders the availability of nutrients for all microorganisms, decreasing the total bacterial count, in addition to yeasts.

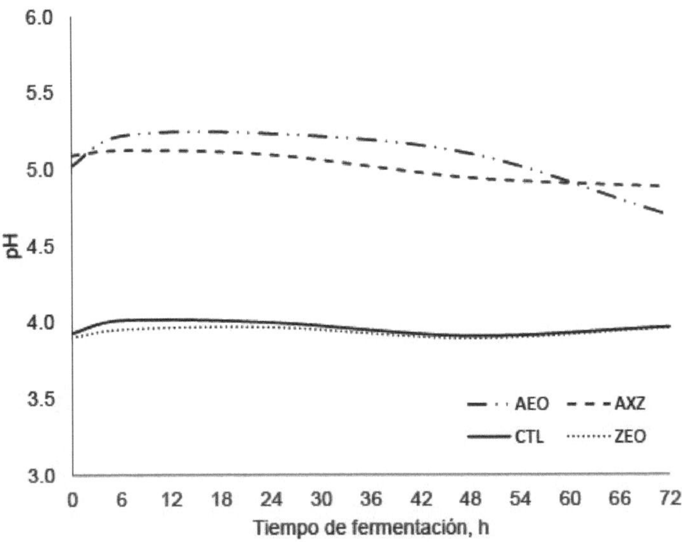

Graph 5. pH in the broth versus time of the experiment.

Table 9. Concentration of total aerobic mesophilic bacteria per mL broth, expressed as CFU/mL broth.

h	Treatment[1]				
	AEO	AXZ	CTL	ZEO	E.E.
6	$2.22\ {\scriptstyle x103}$	1.76×10^2	$4.84\ {\scriptstyle x103}$	1.49×10^2	8.76
24	8.49×10^3	2.25×10^3	1.08×10^4	2.41×10^3	2.04
48	1.74×10^4	9.78×10^3	2.12×10^4	1.05×10^4	4.02
72	1.34×10^4	1.08×10^4	1.64×10^4	1.05×10^4	3.52

[1] AEO indicates treatment with 0.1 % Oregano Essential Oil, ZEO treatment with 2 % Zeolite and AXZ treatment with both (0.1 and 2 % respectively). CTL indicates Control treatment.

CONCLUSIONS AND RECOMMENDATIONS

It is concluded that the use of Oregano Essential Oil provides microbiological advantages over the traditional yeast production process in a molasses and urea broth, thus reducing production time and avoiding contamination by foreign bacteria. These improvements guarantee a clean process free of microbiological contaminants.

The addition of zeolite affects the development of micro-organisms in the broth, therefore, although it has a pH benefit, it is not recommended for use in molasses and urea broths.

Bacterial growth is significantly affected by the inclusion of AEO at low concentrations, which means that with a small amount of AEO, an adequate yeast production process can be guaranteed.

The molasses and urea broth allows the adequate development of the yeast *Kluyveromyces lactis*, as well as its multiplication until reaching acceptable quantities to prepare inoculums for terrestrial animals at a very low cost. Another conclusion is that the inclusion of AEO benefits the process by avoiding the formation of bubbles, thus reducing the loss of broth in the form of foam.

LITERATURE CITED

Abdel-Tawwab, M., A. M. Abdel-Rahman and N. E. M. Ismael. 2008. Evaluation of commercial live bakers' yeast, *Saccharomyces cerevisiae* as a growth and immunity promoter for Fry Nile tilapia, *Oreochromis niloticus* challenged in situ with *Aeromonas hydrophila*. Aquaculture 280:185-189.

Chun, S.-S., D. A. Vattem, Y. T. Lin, and K. Shetty. 2005. Phenolic antioxidants from clonal oregano (*Origanum vulgare*) with antimicrobial activity against Helicobacter pylori. Process Biochem. 40:809-816.

D^az-Plascencia, D. 2011. Development of a yeast-based inoculum and its effect on in vitro fermentation kinetics in rations for high-producing Holstein cows. PhD thesis. Faculty of Zootechnics and Ecology. Autonomous University of Chihuahua. Chihuahua, Chih. Mexico.

Gatesoupe, F. J. 2007. Live yeasts in the gut: natural occurrence, dietary introduction, and their effects on fish health and development. Aquaculture 267:20-30.

Grewal, H. S. and K. L. Kalra. L. Kalra. 1995. Fungal production of citric acid. Biotechnol. Adv. 13:209234.

Krishna, C. 2005. Solid-state fermentation systems-an overview. Crit. Rev. Biotechnol. 25:1-30.

Leung, S., S. Barrington, Y. Wan, X. Zhao and B. El-Husseini. 2007. Zeolite (clinoptilolite) as feed additive to reduce manure mineral content. Bioresour. Technol. 98:3309-3316.

Li, Z. 2003. Use of surfactant-modified zeolite as fertilizer carriers to control nitrate release. Microporous Mesoporous Mater. 61:181-188.

SAS Institute, Inc (2006) SAS/STAT users guide: Statics Version 9 Cary, North Carolina. U.S.A.

Secretary of Health. 1995. Norma oficial mexicana NOM-092-SSA1-1994 Goods and services. Method for counting aerobic bacteria on a plate.

Teixeira, B., A. Marques, C. Ramos, C. Serrano, O. Matos, N. R. Neng, J. M. F. Nogueira, J. A. Saraiva and M. L. Nunes. 2013. Chemical composition and bioactivity of different oregano (*Origanum vulgare*) extracts and essential oil. J. Sci. Food Agric.

- INCLUSION OF FERMENTED APPLE HOPS AND A PROBIOTIC (*Kluyveromyces lactis*) INTO THE DIET OF TILAPIA (*Oreochromis niloticus*) DURING THEIR JUVENILE STAGE

SUMMARY

INCLUSION OF FERMENTED APPLE HOPS AND A PROBIOTIC (*Kluyveromyces lactis*) INTO THE DIET OF TILAPIA (*Oreochromis niloticus*) DURING THEIR YOUTH STAGE

The objective of this work was to evaluate the effects of fermented apple bagasse (BMF) and yeast inoculum (IL) *Kluyveromyces lactis* in the diet of tilapia. The variables evaluated were weight gain, length, feed conversion ratio (FCR) and fillet yield. Four diets were prepared, including 10 % BMF (MAN Diet), 3 % IL (PRB Diet) and a mixture of both (MXP Diet), compared to a control (CTL). Fish were measured weekly and then sacrificed for fillet. Fish fed BMF had a significantly higher slaughter weight (37.93 g) than the control (34.92 g), while those fed IL were significantly lower (32.42 g) ($P < 0.05$). Length gain had a treatment effect ($P > 0.05$), with the MAN diet (92.19 mm) having a greater effect than the control (88.38 mm), and the PRB diet (83.84 mm) ($P < 0.05$). Likewise, there was a difference with respect to AAR ($P > 0.05$) because the MAN diet had an AAR of 1.78, while the control had 1.91. Fillet yield was higher ($P > 0.05$) in the CTL and MAN diet treatments (52.33 and 51.15 %) than MXP and PRB (44.51 and 46.4 %). It is concluded that the BMF acts as a source of microbial protein of good quality and high digestibility, in addition, while the yeast did not show positive effects on the fish, the antioxidant compounds present in the BMF provided benefits that translate into an improvement in the productive parameters.

INTRODUCTION

Tilapia (*Oreochromis* sp.) is a species that has been of increasing interest to the aquaculture industry in recent years, being the second most important freshwater species farmed in the world after carp (Atwood *et al.*, 2003).

In tilapia production systems, a large part of the production costs correspond to feed, using protein of animal origin, generally fishmeal, which considerably increases the cost of feed. Therefore, the search for good quality vegetable or microbial protein sources is a priority in fish production, as they are currently the most demanded source (Ribeiro *et al.*, 2014).

Apples are a natural source of antioxidants, which in addition to providing a large amount of antioxidants, are an excellent substrate for the production of microorganisms through solid-state fermentation, being able to break down cellulosic structures (Dhillon *et al.*, 2012) and release the phenolic compounds contained in them to make them available to consumers of this ferment (Ajila *et al.*, 2011).

Apple bagasse (AB) consists of the residues from the juice production process and is composed of peels, seeds, cores, stalks, pulp remains and soft tissues, and is therefore considered a product of high nutritional value (Garrta *et al.*, 2009). When using this material as an ingredient in animal diets, the producer faces several problems, such as its deficiency of digestible protein, its high moisture content and low pH, so the development of microbial protein in this medium is presented as an excellent alternative to increase its protein level and modify its pH, which can be used for animal feed as a high quality food supplement, providing protein at a lower cost than animal meal, and with a similar digestibility (Bhalla and Joshi, 1994).

Therefore, the objective of this study was to evaluate the effect of the inclusion of Fermented Apple Bagasse (FMB) as a source of protein and antioxidants and the use of yeast inoculum (YI) as a source of probiotics on the production parameters of tilapia.

The study was carried out in the fish farming module of the Facultad de Zootecnia y Ecolog^a of the Universidad Autonoma de Chihuahua, in the city of Chihuahua, Mexico, conditioned in temperature, light and a battery of fish tanks with a recirculating hydraulic system, which includes physicochemical filtering to maintain the biological conditions necessary to sustain the life of aquatic organisms and their optimal development throughout their productive stages.

Water movement was carried out by means of a Pedrollo PKm 60 pump, 0.5 hp, at an average flow rate of 40 l/min for the whole system, thus ensuring adequate oxygenation. The filtration system consists of a 6 mm thick glass tank, which contains filter media composed of activated carbon, granzon and zeolite, which perform physical and chemical filtration, allowing the establishment of denitrifying bacteria colonies, favouring an adequate level of nitrogenous waste compounds in the water. An ultraviolet light system was also installed, which by means of two 15 W tubes, irradiates the water in order to kill possible parasites and their vegetative forms, maintaining the aquaculture health and water quality.

The water quality was constantly monitored in terms of Oxygen Concentration, Nitrate Concentration, CO_2 Concentration, Temperature and Ammonium Concentration, so that they always remained constant and within the optimal range for the species, thus avoiding influencing the results of the experiment.

The yeast inoculum (IL) used was the same as that obtained in Study II, and was adjusted to a concentration of 1×10^6 yeast per mL, by a simple count in a Neubauer chamber by the method described by D^az-Plascencia (Diaz-Plascencia, 2011), where a sample of 1 mL of inoculum is taken, and diluted in serial dilutions, until the appropriate concentration is obtained.

The diet formulation was done in the Nutrion 5 formulation software, based on the N.R.C. (National Research Council) recommendations for Tilapia (*Oreochromis niloticus*), making the necessary modification to include the BMF levels without altering the nutritional aspects of the feed. The diets were formulated as set out in Table 10, using the inclusion level of 10% BMF (MAN), 3% IL (PRO) and their interaction (MXP), contrasted against a control diet (CTL diet). These diets were prepared in the Animal Nutrition laboratory of the Faculty of Animal Husbandry and Ecology, where the ingredients were homogenised and wetted to 24 % w/w in order to be pelletised in a commercial meat mill. The formed pellets were left to dry at room temperature in air permeable trays for 48 hours until they reached a humidity of 8 %, resulting in a yellow coloured feed, with a suitable consistency and a pleasant smell and taste for the fish.

Once the diets were formulated, proximate analysis was carried out by conventional A.O.A.C. (Association of Official Analytical Chemistry) methods to evaluate their nutritional content and to determine if there were any differences between them.

Table 10.- Composition of the diets used in the experiment.

Ingredient	Diet[8]

[8] MAN indicates treatment with 10 % BFM, PRB treatment with 3 % IL and MXP treatment with both (10 and 3 % respectively).

	Control (%)	MAN (%)	PRB (%)	MXP (%)
Soybean paste	62.9	45.8	62.9	45.8
Vegetable oil	6.2	6.4	6.2	6.4
Milk powder	11.0	10.0	11.0	10.0
Ground ma^z	12.5	19.6	12.5	19.6
Meat meal	5.0	5.0	5.0	5.0
Premix V and M	0.3	0.3	0.3	0.3
Ca carbonate	1.1	1.1	1.1	1.1
Salt	1.0	1.0	1.0	1.0
IL	-	-	3.0	3.0
BMF	-	10.0	-	10.0

The fish used were Nile tilapia (*Oreochromis niloticus*) of an average size of one inch, two weeks old, obtained from the fish farm of La Boquilla, in the municipality of San Francisco de Conchos, Chihuahua State, Mexico. The fish obtained were previously masculinised to avoid interference by the sex factor.

Four treatments were established: the diet supplemented with fermented apple bagasse (MAN), the diet supplemented with yeast inoculum (PRO) and their interaction (MXP), as well as the control diet, which was the commercial diet for tilapia (CTL). The following are the treatments:

1. MAN: Diet formulated with 10% BMF.
2. PRO: Diet supplemented with 3% IL.
3. MXP: Diet formulated with 10 % BMF and 3   % IL.
4. CTL: Control diet.

A completely randomised design was used, where glass fish tanks with ten fish each were used as experimental unit, with five replicates for each treatment, giving a total of 20 fish tanks. From each tank, three fish were taken at random each week to obtain the measurements, having three replicates for each repetition. **Response Variables**

Weight and length gain. Over 8 weeks, fish were weighed and measured every 7 d, as described by Crivelenti (Crivelenti and Mundim, 2011), weighed in triplicate on an electronic balance, taken at random from each of the 21 tanks and measured for length, height and width with a millimetre vernier. To facilitate the handling of the animals, once they were removed from their tanks, they were placed in fresh water for a few seconds to reduce their activity and allow them to be handled without risk of damage to the fish themselves.

The diet was fed according to the mass of the experimental unit and at the rate set by the NRC for optimal growth.

Fillet yield. After completion of the trial at 8 weeks, fish were sacrificed by cephalic section after being weighed, measured and ice-sedated. Fillets from each fish were obtained by hand, weighed and frozen at -18 °C until chemical analysis (Fagbenro and Jauncey, 1998).

Survival rate. Fish were counted every d, noting in the logbook if there was mortality every d. In cases where mortality occurred, the body of the fish was analysed for possible signs of disease and recorded in the logbook.

Experimental Design:

For the statistical analysis of the data, a one-way ANOVA test was used to compare the productive variables in each week and proximal analysis, in addition to a Tukey's comparison of means between treatments, considering the treatments statistically significant at $P < 0.05$. For the weight and length gain of the fish with respect to time, a study of repeated means over time was carried out with the Proc Mixed procedure, finding significant effects at $P < 0.05$. The statistical treatment was carried out with the computer package SAS, version 9.1.3 (SAS Institute, 2006).

The proximal analysis of the diets showed that there was only a statistically significant difference in the crude fibre levels, thanks to the inclusion of BMF, which has a significant amount of fibre, however, this inclusion does not modify the energy levels or exceed the recommended amount of fibre for the species (FAO, 2010), so isocaloric and isoproteic diets were used for the experiment (Table 11).

In the present study, a significant effect ($P < 0.05$) of the inclusion of BMF in the diet of growing Tilapia hatchlings was obtained, presenting a higher weight gain ($P < 0.05$), higher length gain ($P < 0.05$) and lower TCA ($P < 0.05$) (Table 12), while the inclusion of IL presented a negative effect on the same variables. On the other hand, the interaction of BMF and IL obtained a good response in the variables studied, although still lower than the diet with BMF, but higher than the control diet (Tables 13 and 14). This effect can be explained by the presence of antioxidant agents in the BMF, which was corroborated in experiment II. This work is in agreement with what has been reported by other authors, where the inclusion of by-products from the fruit industry provides antioxidants to fish diets and improves their productive parameters (Sahin *et al.*, 2014; Brenes *et al.*, 2016).

In the aspect of water quality control and physicochemical parameters of the environment in which the fish developed, it can be said that there was no statistically significant difference between treatments ($P > 0.05$), however, there is a downward trend in temperature due to the normal change in environmental temperature throughout the year, as this experiment was conducted between the months of August and September. Despite this decrease in temperature, the values remained within the appropriate range for the correct development of the species (22 to 32 °C), so that no interference was observed with the experiment (Graph 6), nor by the oxygen concentration, which remained at the appropriate levels as provided by the FAO and SAGARPA technical data sheet for this species (Graph 7).

Regarding waste metabolites, constant monitoring of ammonium and CO_2 showed that at no time were concentrations higher than those recommended for Tilapia (FAO, 2010), with ammonium always below 1 mg/L and CO_2 below 30 mg/L, (Figures 8 and 9). Despite being below the maximum permitted, there was an increase in ammonium concentration in the fourth and fifth week, which can be attributed to the accumulation of waste in the filter, a situation that was corrected by cleaning the filter.

The pH of the water in the tanks varied throughout the experiment, showing that there was no difference between treatments, so it can be said that the inclusion of BMF, IL and their interaction do not modify the pH of the water in the system where the fish develop outside the established limits for this species (6.5 to 8.5), and they behaved in the same way as the control diet (Graph 10).

Mortality in the experiment remained low, with most deaths being caused by fish that jumped out of the tank and were found later, ruling out the presence of disease.

Table 11. Proximate analysis of the diets used in the experiment.

Treatment[1]	Aetheric extract, %	Protema, %	Ash, %	Crude fibre, %, %, %, %,	E.L.N., %

				%, %, %, %, %, %, %, %.	
MAN	8.27[a]	34.31[a]	8.07[a]	8.04[a]	41.30[b]
PRB	8.49[a]	34.98[a]	8.67[a]	5.03[b]	42.80[ab]
MXP	8.77[a]	33.85[a]	7.20[a]	8.16[a]	41.98[b]
Control	8.34[a]	33.51[a]	6.73[a]	5.01[b]	46.39[a]

[1] MAN indicates treatment with 10 % BFM, PRB treatment with 3 % IL and MXP treatment with both (10 and 3 % respectively).

Different literals between rows indicate statistical difference (P > 0.05).

Table 12. Feed conversion ratio (FCR) of fish per treatment[1] during the experiment [9] [10]

Week	CTL	MAN	MXP	PRB
1	2.3469 [ab]	2.1356 [b]	2.0791 [b]	2.8724 [a]
2	1.7044 [ab]	1.2061 [b]	1.8873 [ab]	1.9057 [a]
3	1.8944 [a]	1.6103 [ab]	1.4577 [b]	1.7866 [a]
4	1.8958 [a]	1.0950 [d]	1.5841 [c]	1.7500 [b]
5	1.5593 [c]	2.2320 [a]	1.6496 [b]	2.0565 [ab]
6	2.0058 [c]	1.9989 [c]	2.4841 [b]	2.9477 [a]
7	1.8993 [a]	1.6871 [b]	1.8531 [ab]	1.9200 [a]
8	2.0392 [c]	2.3483 [ab]	2.3062 [ab]	2.4100 [a]
Average	1.9181 [ab]	1.7891 [c]	1.9127 [ab]	2.2061 [a]

Table 13. Average weekly weight in grams of fish per treatment[1] during the experiment. [11] [12]

Week	CTL (g)	MAN (g)	MXP (g)	PRB (g)	E.E.
0	7.14[a]	7.5[a]	8.15[a]	6.87a	0.50
1	8.30[ab]	9.46[a]	9.08[ab]	7.21a	0.52
2	10.54[ab]	12.65[a]	11.03[ab]	10.06[b]	0.67
3	12.76[b]	15.78a	14.05[ab]	12.30[b]	0.80
4	17.69[ab]	21.34a	18.26[b]	17.16[b]	1.05
5	21.5[ab]	24.75a	22.22[ab]	20.14[b]	1.28
6	25.32[ab]	27.7[a]	25.41[ab]	22.58[c]	1.50
7	30.25[ab]	33.44[a]	30.21[ab]	27.77[c]	1.57
8	34.92[ab]	37.93[a]	34.34[ab]	32.42c	1.80

Table 14. Average weekly length in miKmetres of fish per treatment[1] during the experiment. [13] [14]

---

[9] MAN indicates treatment with 10 % BFM, PRB treatment with 3 % IL and MXP treatment with both (10 and 3 % respectively).
[10] Different literals between columns indicate statistically significant difference in Tukey's comparison of means (P > 0.05).
[11] MAN indicates treatment with 10 % BFM, PRB treatment with 3 % IL and MXP treatment with both (10 and 3 % respectively).
[12] Different literals between columns indicate statistically significant difference in Tukey's comparison of means (P > 0.05).
[13] MAN indicates treatment with 10 % BFM, PRB treatment with 3 % IL and MXP treatment with both (10 and 3 % respectively).

Week	CTL (mm)	MAN (mm)	MXP (mm)	PRB (mm)	E.E.
0	48.13a	48.90a	49.68a	48.72a	2.17
1	52.62^a	56.08^b	52.90a	52.00^a	2.25
2	56.90^{ab}	60.86^a	58.29^{ab}	55.44^b	2.13
3	62.91^{ab}	65.07^a	61.81^{ab}	58.65^b	2.18
4	68.03^{ab}	71.38^a	67.82^{ab}	65.26^b	2.50
5	72.68^{ab}	77.84^a	73.06^{ab}	70.48^b	2.48
6	76.73^{ab}	81.37^a	76.14^{ab}	75.08^b	2.26
7	82.60^{ab}	86.51^a	82.86^{ab}	80.28^b	2.53
8	88.38^{ab}	92.19^a	89.51^{ab}	83.84^c	2.64

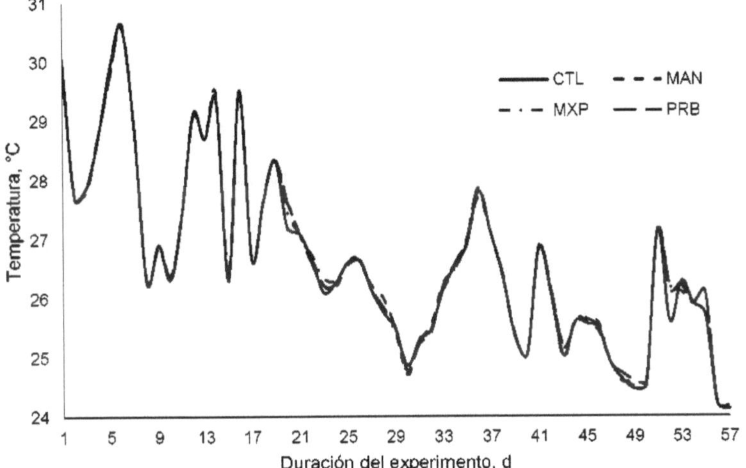

Graph 6. Temperature of the water in the fish tank system with respect to the time of the experiment.

[14] Different literals between columns indicate statistically significant difference in Tukey's comparison of means (P > 0.05).

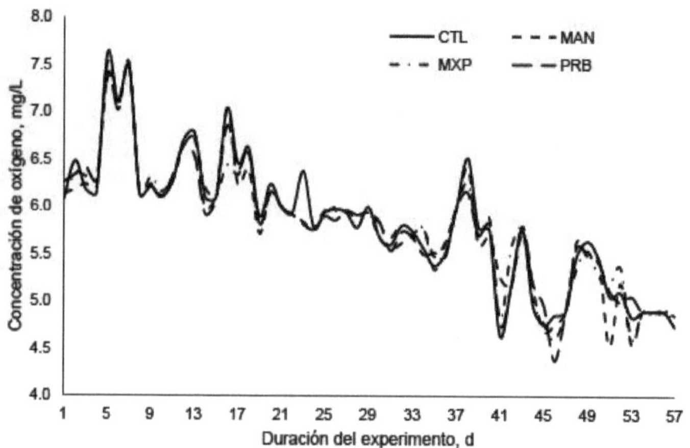

Graph 7. Concentration of oxygen in the water of the fish tank system with respect to the time of the experiment.

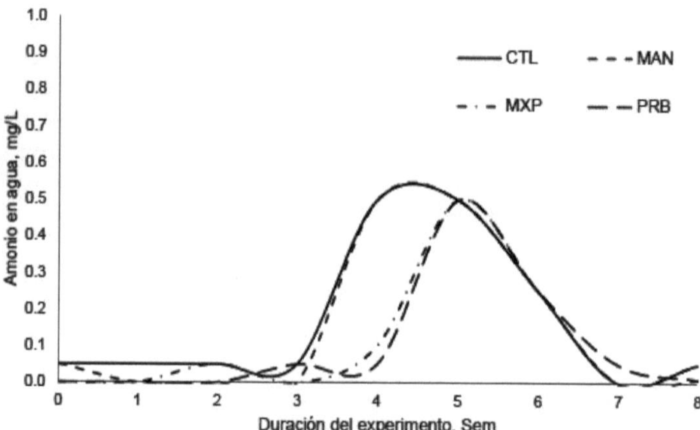

Graph 8. Ammonium concentration in the water of the fish tank system with respect to the time of the experiment.

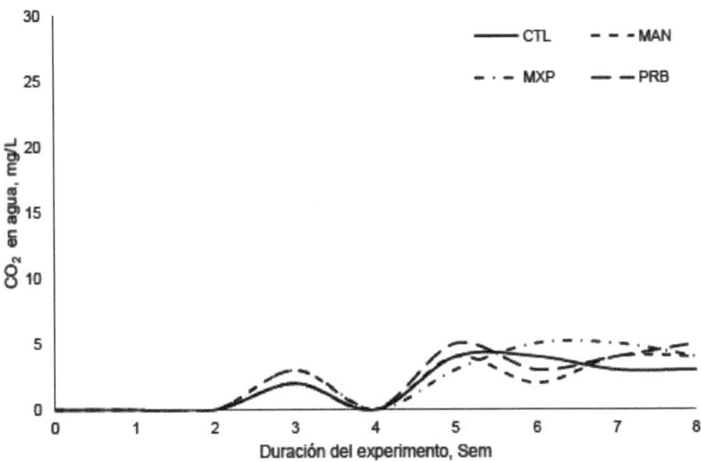

Graph 9. CO_2 concentration in the water of the fish tank system with respect to the time of the experiment.

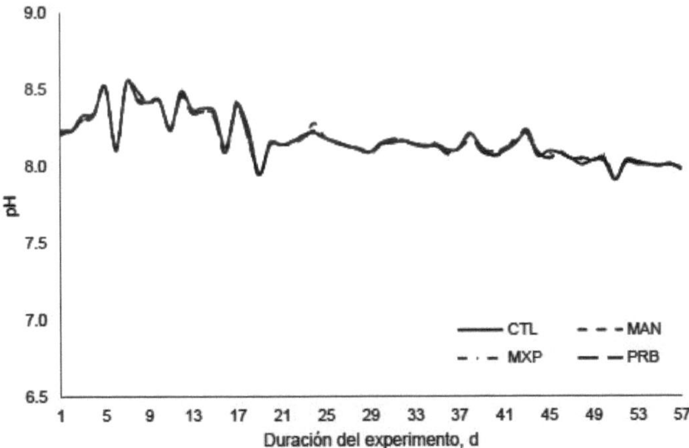

Graph 10. pH of the water in the fish tank system with respect to the time of the experiment.

or bacterial parasites to the revision of the dead fish. Survival remained between 92 and 100 % per treatment (Table 15). This survival rate is adequate for the species, as it ranges between 90 and 100 %, according to FAO and SAGARPA technical data sheets, and is similar to that found by other authors for tilapia growth (Reque *et al.*, 2010).

Regarding the fillet yield parameters, it was determined that there were statistically significant differences between the treatments, i.e. there was an effect produced by the inclusion of BFM and IL in the diets (Table 16). The diet that produced the best fillet yield was the control diet (P > 0.05), with 52.33 % fillet per fresh weight, followed by the MAN diet, which had a fillet yield of 51.15 %, very close to the control, while the MXP and PRB diets were considerably different from the first two, with fillet yields of 44.51 % and 46.44 % respectively. These results are in agreement with those presented by Rojas *et al.* in 2011, where morphometric measurements agree for carcass weight, length and fillet yield (Rojas-runjaic *et al.*, 2011).

This effect can be explained by the inclusion of BFM, which provides antioxidants to the fish, so that it decreases oxidative stress and allows it to produce more muscle mass, because fewer free radicals cause less damage to muscle tissues and promote muscle development (Dhillon *et al.*, 2013).

Table 15. Survival of fish per treatment during the entire experiment

Treatment[1]	Initial Fish	Fish at the End	Survival
1.- Control	50	46	92 %
2.- MAN	50	49	98 %
3.- MXP	50	50	100 %
4.- PRB	50	47	94 %

[1] MAN indicates treatment with 10 % BFM, PRB treatment with 3 % IL and MXP treatment with both (10 and 3 % respectively).

Table 16. Average fillet yield of fish by treatment at slaughter

Treatment[1]	Fillet yield, % Fillet yield, %, %, %, %, %, %, %, %, %, %. [15][16]	D.E.
CTL	52.33[a]	3.83
MAN	51.15[ab]	3.84
MXP	44.51[c]	6.06
PRB	46.44[bc]	2.71

<hr>

[15] MAN indicates treatment with 10 % BFM, PRB treatment with 3 % IL and MXP treatment with both (10 and 3 % respectively).
[16] Different literals between columns indicate statistically significant difference in Tukey's comparison of means (P > 0.05).

CONCLUSIONS AND RECOMMENDATIONS

This study concludes that aquaculture is an excellent alternative for the production of animal protein in an accessible, safe and environmentally friendly way. However, the main cost involved in the management of a fish farm is the feed, which obliges the producer to look for a feed that presents an adequate balance between cost and quality.

It is also concluded that in order to develop a fish feed that is in an adequate cost range it is necessary to use alternative sources of protein, and a suitable alternative to supply high cost protein is the addition of fermented apple by-products in solid form.

The yeasts contained in the inoculum produced in the fermentation were unable to act as probiotics in the fish, failed to remain in the digestive tract of the fish, and did not provide any digestive benefit, but rather had a statistically negative influence on fish development, preventing the fish from reaching their full potential, so the use of *Kluyveromices lactis* yeast in the diet of Tilapia fish is not recommended.

It is concluded that apple bagasse has antioxidant compounds, as apples are rich in them, and these will serve to prevent oxidative stress in the fish by free oxygen radicals, improving their productive performance in terms of weight gain, growth in length, feed conversion rate and fillet yield.

It is concluded that the use of BFM provides antioxidant compounds of the polyphenol family in significant amounts, which improve the productive parameters of the fish, obtaining a better weight gain, length growth, feed conversion rate and fillet yield, while the inclusion of IL does not produce a positive difference with respect to the control. Therefore, it can be said that the yeasts *Kluyveromyces lactis* used in the inoculum show a significant negative effect on production parameters.

The inclusion of BMF in the diet of tilapia is therefore a good way to tackle the contamination caused by apple by-products generated in the apple processing industries and to put them to productive use in aquaculture.

LITERATURE CITED

Ajila, C. M., F. Gassara, S. K. Brar, M. Verma, R. D. Tyagi and J. R. Valero. 2011. Polyphenolic antioxidant mobilization in apple pomace by different methods of solid-State fermentation and evaluation of Its antioxidant activity. Food Bioprocess Technol. 5:2697-2707.

Atwood, H. L., J. R. Tomasso, K. Webb and D. M. Gatlin. 2003. Low-temperature tolerance of Nile tilapia, *Oreochromis niloticus*: effects of environmental and dietary factors. Aquac. Res. 34:241-251.

Bhalla, T. C. and M. Joshi. 1994. Protein enrichment of apple pomace by co-culture of cellulolytic moulds and yeasts. World J. Microbiol. Biotechnol. 10:116-7.

Brenes, A., A. Viveros, S. Chamorro and I. Arija. 2016. Use of polyphenol-rich grape byproducts in monogastric nutrition. A review. Anim. Feed Sci. Technol. 211:1-17.

Crivelenti, L. Z. and A. V Mundim. 2011. Serum biochemical values of nile tilapia (*Oreochromis niloticus*) in intensive culture. Rev. Investig. Vet. Peru 22:318-323.

Dhillon, G. S., S. Kaur, S. K. Brar and M. Verma. 2012. Potential of apple pomace as a solid substrate for fungal cellulase and hemicellulase bioproduction through solid-state fermentation. Ind. Crops Prod. 38:6-13.

Dhillon, G. S., S. Kaur and S. K. Brar. 2013. Perspective of apple processing wastes as lowcost substrates for bioproduction of high value products: A review. Renew. Sustain. Energy Rev. 27:789-805.

D^az-Plascencia, D. 2011. Development of a yeast-based inoculum and its effect on in vitro fermentation kinetics in rations for high-producing Holstein cows. PhD thesis. Faculty of Zootechnics and Ecology. Autonomous University of Chihuahua. Chihuahua, Chih. Mexico.

Fagbenro, O. and K. Jauncey. 1998. Physical and nutritional properties of moist fermented fish silage pellets as a protein supplement for tilapia (*Oreochromis niloticus*). Anim. Feed Sci. Technol. 71:11-18.

FAO. 2010. World aquaculture 2010. 1st ed. (FAO Fisheries and Aquaculture Department, editor.). Rome.

Garrta, Y. D., B. S. Valles and A. P. Lobo. 2009. Phenolic and antioxidant composition of byproducts from the cider industry: Apple pomace. Food Chem. 117:731-738.

Reque, V. R. R., J. R. E. de Moraes, M. A. de Andrade Belo and F. R. de Moraes. 2010. Inflammation induced by inactivated *Aeromonas hydrophila* in Nile tilapia fed diets supplemented with *Saccharomyces cerevisiae*. Aquaculture 300:37-42.

Ribeiro, C. S., R. G. Moreira, O. a. Cantelmo and E. Esposito. 2014. The use of *Kluyveromyces marxianus* in the diet of red-stirling tilapia (*Oreochromis niloticus*, Linnaeus) exposed to natural climatic variation: Effects on growth performance, fatty acids, and protein.
deposition. Aquac. Res. 45:812-827.

Rojas-runjaic, B., D. A. Perdomo and D. E. Garrta. 2011. Carcass yield and filleting of tilapia (*Oreochromis niloticus*) Chitralada variety produced in Trujillo State, Venezuela. Zootec. Trop. 29:113-126.

SAGARPA and CONAPESCA. 2012. Anuario estadistico de acuacultura y pesca 2011.

Sahin, K., C. Orhan, H. Yazlak, M. Tuzcu and N. Sahin. 2014. Lycopene improves activation of antioxidant system and Nrf2/HO-1 pathway of muscle in rainbow trout (*Oncorhynchus mykiss*) with different stocking densities. Aquaculture 430:133-138.

SAS Institute, Inc. 2006. SAS/STAT users guide: Statics Version 9 Cary, North Carolina. U.S.A.

Secretary of Health. 1995. Norma oficial mexicana NOM-092-SSA1-1994 Goods and services. Method for counting aerobic bacteria on a plate.

yes
I want morebooks!

Buy your books fast and straightforward online - at one of world's fastest growing online book stores! Environmentally sound due to Print-on-Demand technologies.

Buy your books online at
www.morebooks.shop

Kaufen Sie Ihre Bücher schnell und unkompliziert online – auf einer der am schnellsten wachsenden Buchhandelsplattformen weltweit! Dank Print-On-Demand umwelt- und ressourcenschonend produziert.

Bücher schneller online kaufen
www.morebooks.shop

info@omniscriptum.com
www.omniscriptum.com

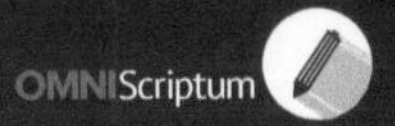

Printed by Books on Demand GmbH, Norderstedt / Germany